Case Studies in Fisheries Conservation and Management:

Applied Critical Thinking & Problem Solving

Authors' proceeds from these case-study textbooks will be placed into the John E. Skinner Memorial Travel Award endowment to support student travel to annual American Fisheries Society meetings

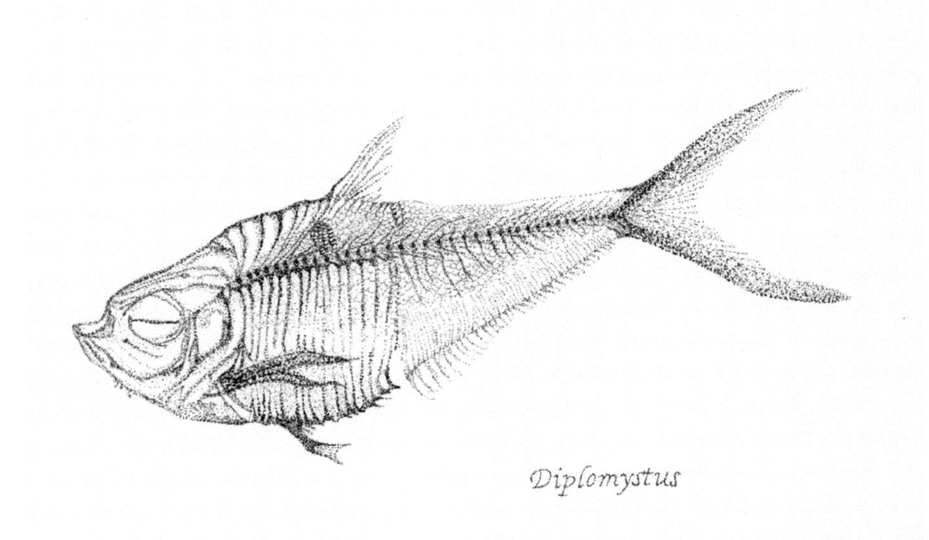

Diplomystus

Case Studies in Fisheries Conservation and Management:
Applied Critical Thinking & Problem Solving

Brian R. Murphy
Department of Fisheries and Wildlife Sciences, Virginia Tech
Blacksburg, Virginia 24061, USA

David W. Willis
Department of Wildlife and Fisheries Sciences, South Dakota State University
Brookings, South Dakota 57007, USA

Michelle D. Klopfer
Department of Fisheries and Wildlife Sciences, Virginia Tech
Blacksburg, Virginia 24061, USA

Brian D. S. Graeb
Department of Wildlife and Fisheries Sciences, South Dakota State University
Brookings, South Dakota 57007, USA

American Fisheries Society
Bethesda, Maryland
2010

A suggested citation format for this book follows.

Murphy, B. R., D. W. Willis, M. D. Klopfer, and B. D. S. Graeb. 2010. Case studies in fisheries conservation and management: applied critical thinking and problem solving. American Fisheries Society, Bethesda, Maryland.

Printed in the United States of America on acid-free paper.

Library of Congress Control Number 2010934391
ISBN 978-1-934874-18-9

American Fisheries Society Web site address: *www.fisheries.org*

American Fisheries Society
5410 Grosvenor Lane, Suite 100
Bethesda, Maryland 20814
USA

Table of Contents

Acknowledgments

Foremost, the authors gratefully acknowledge the Acorn Alcinda Foundation of Lewes, Delaware. Without their support, this book never could have been completed. The Foundation supported everything from international travel that led directly to several of these case studies, to extensive editorial assistance that made completing this unique book possible. Kit and Jan Kennedy of the Foundation have been particularly supportive of our efforts to improve university education, which led directly to their critical support of this book.

Editorial production of this book was also partially supported by the College of Agriculture and Biological Sciences at South Dakota State University.

The authors are very grateful for the tireless work of editorial assistant Gloria Schoenholtz. She kept us on track and pushed us to complete critical tasks when we all had other professional distractions, and she managed to pull together the disparate contributions of multiple authors into a coherent whole. Thank you sincerely, Gloria.

The original idea for this book, and the first case herein, grew from Brian Murphy's involvement in The Global Seminar (www.globalseminar.org), a distance learning consortium of universities around the world addressing education in sustainability; he gratefully acknowledges guidance from Global Seminar colleagues, particularly Tom Hammett, Jim McKenna, Dean Sutphin, and Louise Buck. Brian Murphy and David Willis also attended a case-teaching workshop sponsored by the National Center for Case Study Teaching in Science (NCCSTS), which directly influenced our case teaching and then our writing for this book; we extend our thanks to NCCSTS Director Kipp Herried for that opportunity.

We gratefully acknowledge scientists from a multitude of agencies and institutions who provided data and other supporting materials that are the heart of various cases in this book: Case 3—Craig Milewski (Paul Smith's College); Case 4—Larry Goedde (Ohio Department of Natural Resources) and Daniel Coble (Wisconsin Cooperative Fishery Research Unit [retired]); Case 6—Craig Paukert (Missouri Cooperative Fish and Wildlife Research Unit); Case 8—T.J. DeBates (Minnesota Department of Natural Resources) and Craig Paukert; Case 9—Tim Bister (Texas Parks and Wildlife Department) and Todd St. Sauver (South Dakota Department of Game, Fish and Parks); Case 11—Bob Wiley (Wyoming Game and Fish Department [retired]); Case 12—John Lindgren (Minnesota Department of Natural Resources), Robert Neumann (In-Fisherman, Inc.), Carter Kruse (Turner Foundation); Case 14—John Lott (South Dakota Department of Game, Fish and Parks) and Trent Sutton (University of Alaska-Fairbanks); Case 16—Brian Blackwell (South Dakota Department of Game, Fish and Parks) and Stephen Wilson (National Park Service); Case 19—Dane Shuman and Steve Krentz (U.S. Fish and Wildlife Service); Case 20—Charlie Munger and Brian Van Zee (Texas Parks and Wildlife Department); Case 21—John Lott and Robert L. Hanten (South Dakota Department of Game, Fish and Parks); Case 22—Steve Eder (Missouri Department of Conservation); Case 26—Nelson Beretta (Empresa de Pesquisa Agropecuária e Extensão Rural de Santa Catarina, Brasil), José Dotta and Adil Vaz (Universidade Estadual de Santa Catarina, Lages, Brasil); Case 28—Patrick Martinez (Colorado Division of Wildlife); Case 31—Don Gabelhouse, Jr., Brad Newcomb, and Peter Spirk (Nebraska Game and Parks Commission).

We thank many scientific colleagues who provided technical reviews that helped to improve these case studies: Geno Adams, José Luis Arreguin, Katie Bertrand, Louise Buck, Vic DiCenzo, Monique Dufour, Stephen Flickinger, Mark Fritz, Larissa Graham, Chris Guy, Mike Hansen, Helen He, Samara Hermes-Silva, Steve Kelsch, Nick LaPointe, Thomas Lauer, Robert Leaf, Craig Milewski, Robert Neumann, James Parkhurst, Craig Paukert, Kevin Pope, Michael Prager, Brad Ray, Sergio Rodriguez, Chuck Scalet, Trent Sutton, Justin VanDeHey, Amy Villamagna, Bob Wiley, Melissa Wuellner, Songguang Xie, and Evoy Zaniboni-Filho.

We thank Noah Schoenholtz for the cover artwork and designs; and Diane Brown (page ii and Case 13) and Philip Davis (Cases 1, 23, and 28) for donating the use of their original artwork.We acknowledge NOAA for providing public access to many of the photographs used in this book. Also, we thank the Maine Department of Marine Resources Recreational Fisheries program and the Maine Outdoor Heritage Fund for use of many of the fish illustrations found in this book and the accompanying slide sets. Other photographs and figures were graciously provided by various people, agencies, and publications, as listed in **Figure Credits**.

List of Scientific Names

Common and scientific names used in this book

Fishes

American eel........*Anguilla rostrata*
Atlantic bluefin tuna........*Thunnus thynnus*
Atlantic cod........*Gadus morhua*
Atlantic menhaden........*Brevoortia tyrannus*
Atlantic salmon........*Salmo salar*
Black bullhead........*Ameiurus melas*
Black crappie........*Pomoxis nigromaculatus*
Black sea bass........*Centropristis striata*
Blue catfish........*Ictalurus furcatus*
Bluegill........*Lepomis macrochirus*
Blue tilapia........*Oreochromis aureus*
Bonytail........*Gila elegans*
Bull trout........*Salvelinus confluentus*
Channel catfish........*Ictalurus punctatus*
Chilean sea bass........*Dissostichus eleginoides*
Colorado pikeminnow........*Ptychocheilus lucius*
Common carp........*Cyprinus carpio*
Coney........*Cephalopholis fulva*
Crucian carp........*Carassius carassius*
Cutthroat trout........*Oncorhynchus clarkii*
    Colorado River cutthroat trout........*Oncorhynchus clarkii pleuriticus*
    Snake River cutthroat trout........*Onchorhynchus clarkii behnkei*
    Westslope cutthroat trout........*Oncorhynchus clarkii lewisi*
    Yellowstone cutthroat trout........*Oncorhynchus clariki bouvieri*
Dogfish (or Saicanaga)........*Oligosarcus* spp.
Dourado........*Salminus brasiliensis*
Fathead minnow........*Pimephales promelas*
Flannelmouth sucker........*Catostomus latipinnis*
Flathead catfish........*Pylodictis olivaris*
Flathead chub........*Platygobio gracilis*
Florida largemouth bass........*Micropterus s. floridanus*
Gizzard shad........*Dorosoma cepedianum*
Gobies........Gobiidae
Golden shiner........*Notemigonus crysoleucas*
Goldfish........*Carrassius auratus*

Green sunfish........*Lepomis cyanellus*
Humpback chub........*Gila cypha*
Inland silverside........*Menidia beryllina*
Jundiá........*Rhamdia quelen*
Kokanee salmon........*Oncorhynchus nerka*
Lake trout........*Salvelinus namaycush*
Lambari........*Mimagoniates* spp. and *Astyanax* spp.
Largemouth bass........*Micropterus salmoides*
Longear sunfish........*Lepomis megalotis*
Northern largemouth bass........*Micropterus s. salmoides*
Northern pike........*Esox lucius*
Orange roughy........*Hoplostethus atlanticus*
Pacific sardine........*Sardinops sagax*
Pacific viperfish........*Chauliodus macouni*
Pacus........Serrasalminae
Pallid sturgeon........*Scaphirhynchus albus*
Piranhas........*Serrasalmus* spp. and *Pygocentrus* spp.
Pumpkinseed........*Lepomis gibbosus*
Rainbow smelt........*Osmerus mordax*
Rainbow trout........*Oncorhynchus mykiss*
Razorback sucker........*Xyrauchen texanus*
River carpsucker........*Carpiodes carpio*
Roundtail chub........*Gila robusta*
Smallmouth bass........*Micropterus dolomieu*
Sorubim........*Pseudoplatystoma* spp.
Striped bass........*Morone saxatilis*
Swordfish........*Xiphias gladius*
Tambica........*Oligosarcus* spp.
Traira........*Hoplias* spp.
Tilapias........*Oreochromis* spp.
Walleye........*Sander vitreus*
Warmouth........*Lepomis gulosus*
White bass........*Morone chrysops*
White crappie........*Pomoxis annularis*
White sucker........*Catostomus commersonii*
Yellow perch........*Perca flavescens*

Invertebrates

Calico crayfish........*Orconectes immunis*
Channeled whelk........*Busycon canaliculatum*
Horseshoe crab........*Limulus polyphemus*
Opossum shrimp........*Mysis relicta*

Mammals
California sea lion..*Zalophus californianus*

Birds
Red knot..*Calidris canutus*
Black-necked crane...*Grus nigricollis*

Reptiles
Loggerhead sea turtle..*Caretta caretta*

Amphibians
Bullfrog...*Rana catesbeiana*

Plants
Araucaria (or Parana) pine..*Araucaria angustifolia*
Bulrush..*Scirpus* spp.
Cattail..*Typha* spp.
Eurasian watermilfoil...*Myriophyllum spicatum*
Hydrilla..*Hydrilla verticillata*

Introduction

A Message to Students: This Book Can Help You Get a Job

Computer models released last week by scientists at the National Center for Atmospheric Research indicated that the oil from the spill could foul thousands of miles of the Atlantic coast as early as this summer.

—*USA Today, 9 June 2010 (pp. 1–2A)*

A Crisis of Monumental Proportions

As I write this message, hundreds of thousands of gallons of crude oil continue to spill daily almost unfettered into the Gulf of Mexico from a damaged deep-sea well. It is already the largest spill in U.S. history, and there is no end in sight. Oil has begun washing up on the beaches of Florida, and could possibly reach as far as North Carolina if the oil enters the currents that sweep around the southern tip of Florida and join the Atlantic Gulf Stream. Oil has invaded the fragile marshlands of coastal Louisiana, and the marsh vegetation is dying. People in Louisiana are so frustrated at the slow pace of recovery efforts that they have begun sucking up oil from the marsh with common shop vacuums. The first tropical storm of the season is headed across the Gulf towards Texas, and storm-driven tides will carry oil many miles further into these marshlands in the next few days. The oil will directly impact countless wildlife species in the marsh and significantly reduce the recruitment of many important commercial and sport fishes in the Gulf that use the marshes and other coastal areas as breeding or nursery grounds. Birds, marine mammals, sea turtles, and other wildlife are being trapped and fouled by the oil, and many will die uncounted. Oyster reefs have been killed by oil fouling. The National Oceanic and Atmospheric Administration (NOAA) has already deployed a research vessel to the Gulf of Mexico, to begin assessment of the impacts of the spill on sensitive stocks of tunas, swordfish, and sharks (http://www.noaanews.noaa.gov/stories2010/20100625_delaware.html). More than one-third of the U.S. federal waters in the Gulf have been closed to commercial and sport fishing. These closures are devastating the livelihoods of thousands of people. Many private sport fishing trips, charters, and public events such as billfish tournaments have been canceled, which has directly impacted the tourism industry. Commercial fishing currently accounts for some $15 billion in annual economic activity in the Gulf, and recreational fishing adds more than another $2 billion. The spill will undoubtedly cause billions of dollars in economic losses to commercial fishers, charter boats, and countless coastal businesses such as restaurants, hotels, tackle and bait suppliers, fuel suppliers, marinas, fish processors, and numerous others. When you consider non-fishing related tourism in the Gulf, the numbers become even more staggering. Birdwatchers alone account for more than $7 billion in economic activity in the region. Annual recreation-related expenditures in coastal communities along the Gulf may top $34 billion (http://docs.nrdc.org/water/files/wat_10051101a.pdf). It is clear that the Gulf ecosystem is a major economic resource for the region and the nation as a whole, and all of it is in jeopardy. The effects of this spill are likely to be felt in the Gulf and beyond for decades to come, and very well could influence you on a professional and personal level for much of your career.

People everywhere are asking questions about how we got into this difficult situation, and what should be done at this point about these crucial issues and problems.

- Why has the flow of oil continued for more than 2 months now?
- Why is oil continuing to foul new coastline almost every day, despite the fact that we knew for weeks that it was coming?
- Why has the response been so slow, and so seemingly ineffective?
- Given the scale of the potential physical, biological, and economic damage, why were there no contingency plans in place for immediate response, just in case such an accident occurred?
- What will be the short-term and long-term effects on fisheries in the Gulf, and on people who rely on these fisheries for their livelihoods?
- What is the appropriate response to this widespread crisis, by the federal and state agencies charged with the protection, monitoring, and management of these fisheries and other aquatic resources in the Gulf?
- What other fisheries might be impacted if the plume of oil finds its way around the southern tip of Florida, and enters the open waters of the Atlantic Ocean? What other countries will be impacted?
- What contingency plans should be under development right now, for everything from endangered species to important commercial fisheries to critical ecosystem components that might be threatened by the oil?

- What are the potential long-term impacts on the aquatic food web in the Gulf?
- What research needs to be done now in anticipation of coming problems, and later to monitor the effects of the spill?
- How long will the spill impact the wide spectrum of Gulf-related commerce in the region, particularly the overwhelmingly important tourism sector?

By the time that you read this perhaps the well will have been capped, some of these questions will have been answered, and clean-up efforts will be further along than now. But, even so, you should still wonder why this happened in the first place, and why professionals were so slow to respond to this monumental crisis in a timely and effective manner? The list of questions, issues, and problems related to this environmental catastrophe is endless. But who is going to address these issues and problems? Do you believe that you have the skills that will be required to effectively address these problems, which may well be the primary environmental issue for most of your upcoming professional career?

You may say that you are studying to be a freshwater fisheries ecologist or manager rather than someone planning to work in the realm of marine resources, or that you live nowhere near the Gulf and the problem is distant from your personal and professional world. If so, you are deceiving yourself. Ecologist Barry Commoner once said, "The first law of ecology is that everything is connected to everything else. There is one ecosphere for all living organisms and what affects one, affects all." (http://en.wikipedia.org/wiki/Barry_Commoner). You likely have eaten shrimp from the Gulf, or one of the various types of snappers, groupers, and flounders that make up the popular sport and commercial catch. Have you ever eaten swordfish or bluefin tuna? Both spawn in the Gulf of Mexico, and the Gulf and U.S. Atlantic coastal waters are critical juvenile rearing areas. The Gulf is the sole spawning ground for the western stock of Atlantic bluefin tuna, and this oil spill may very well result in this already severely depleted fish stock being listed as officially endangered (http://www.nytimes.com/2010/06/24/us/24fish.html?th=&adxnnl=1&emc=th&adxnnlx=1277377970-gwpOnW0OTtj3TV15ZPaskw&pagewanted=print). Birds in your area may migrate to or through the impacted areas. Colleagues from my department are already deploying to the Gulf coast to monitor the arrival (and hopefully the survival) of endangered shorebirds that recently bred farther north and are currently migrating to the very beach areas being impacted by oil. Your geographic area may very well receive people who decide to leave the Gulf coast area to seek employment elsewhere, and your tax dollars will be committed for decades to various efforts to assess, clean up, and mitigate the effects of this oil spill. Even if you live in a country remote from the spill, you may be affected by global changes in oil production methods that will affect global oil prices and eventually your own pocketbook.

Many people argue that much of this ecological damage could have been prevented by better planning before the disaster and more-effective response after the explosion that started the spill. Have you been developing the skills that will be required to address these vast environmental challenges? Will you be ready for the next regional or global environmental crisis? What skills are most important for you to become an effective natural resources professional in today's world? I contend that it is not the "information" that you may have memorized from various textbooks and courses; you likely did not retain that information for very long following your exams anyway. I contend that your critical-thinking and problem-solving skills are most crucial for you to develop during your college career. You can memorize scientific names and how to identify species, and you can memorize and regurgitate various facts on exams that may ensure that you receive good grades in your courses, but those things are not going to make you an effective professional. Critical thinking and effective problem solving seem to be in short supply as our society attempts to respond to the present Gulf crisis. Are your critical-thinking and problem-solving skills as sharp as they will need to be to address the next unique crisis—and to get you that job?

Thinking and Learning

You likely have taken several years of college courses in natural resources and other disciplines. Undoubtedly you remember a lot of (or at least some!) information from these courses, but has memorizing that information improved your thinking skills? Both our physical and our mental capabilities improve when we face and conquer challenges at the very limits of our present abilities. What types of thinking challenges have you been asked to confront? Psychologists classify learning, and the various types of thinking related to learning, as shown in Figure 1. What types of thinking have you employed in the courses you have completed? Undoubtedly you have been asked many times to operate at the levels of "remember," and "understand." These thinking types are commonly required on multiple choice exams, where you are tested on your retention of a body of information to which you were exposed in lectures or a textbook. You also probably had some courses that required the third level of thinking, by "applying" something that you learned (a classification scheme, a formula to solve a problem, etc.). But today can you still apply everything that you "learned" in basic chemistry, physics, or calculus? Not if you are like me when I was an undergraduate! Why do we forget what we have "learned?" Because learning is not simply memorizing, or not even just

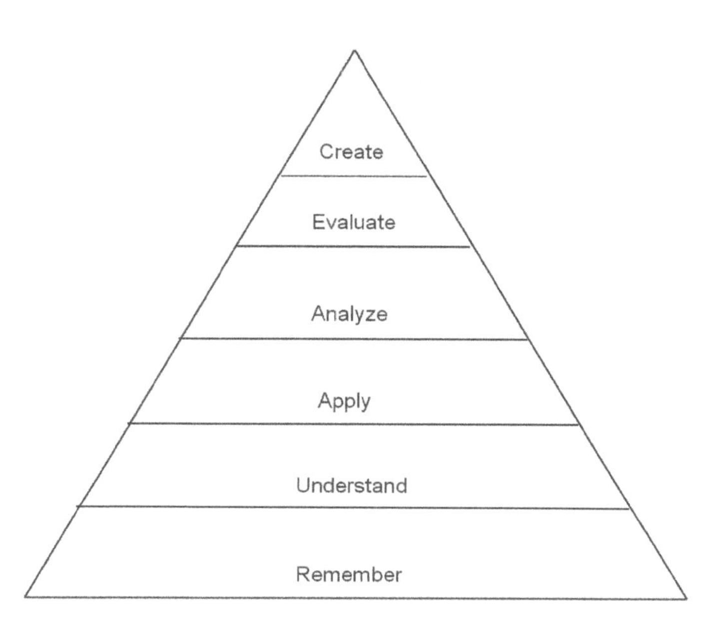

Figure 1. Progressively complex cognitive activities required for students to create or enhance their knowledge base. This classification scheme is commonly known as "Bloom's Taxonomy" (adapted from Anderson and Krathwohl 2001).

understanding it well enough to apply it. Deep learning, as some call it, occurs when we operate at the top three levels of Bloom's Taxonomy (Figure 1). The critical and creative thinking that is required at these levels allows us to do meaningful things with the information that we accumulated in the lower levels. These higher-order thinking tasks help us to organize and access information in new ways, and reveal insights that we never had when we were simply trying to remember topics for an exam. Deep thinking allows us to construct our own personally organized knowledge base in a way that we can access and use it to help us solve problems that we will face throughout our professional (and personal) lives. Complex problems like those we now face in the Gulf of Mexico are not going to be solved with memorized information. They will only be solved by professionals who can effectively analyze problems; evaluate complex situations; and create, evaluate, and apply unique ideas to solve those problems. Real-life natural resource problems demand critical thinking and problem solving, not simply "remembering" and "applying."

What Makes a Fisheries Professional Effective?

Perhaps a good way to understand the crucial skills that you will need as a fisheries professional is to ask yourself the following questions:

- Can you look at a problem or question and design a study, including an effective and efficient sampling regime, to answer the pertinent questions?
- Can you look at a data set and logically postulate the effect of bias in methods on the resultant data?
- Can you look at a data set collected by you or by others and interpret the full meaning of the data, and use the data to support or refute various hypotheses?
- Can you look at a study and critically evaluate the results as to whether they truly answer the question at hand? Can you look at the methods used and determine whether they were appropriate for the hypotheses being tested?
- Can you evaluate the conclusions that others may have drawn from a study and determine whether they are justified by the methods employed?
- Can you effectively and efficiently communicate the results of a study to other scientists, including justifying the methods and the conclusions? Can you communicate the results of the same study to laypersons, in terms that they can understand?
- Can you look at a fisheries conservation or management problem and design a plan to accomplish a set of given objectives?
- Can you create and justify a fishery conservation or management plan to optimally balance competing stakeholder interests?

The critical-thinking, creative-thinking, and problem-solving skills necessary to perform the professional tasks listed above are the very skills that this book is designed to help you develop.

How to Learn from the Cases in this Book

The case studies in this book are real. All of them are situations that we (the authors) have faced, or they are composites of challenges that actual fisheries professionals have faced or are facing now. These case studies are not "case histories" that simply describe the history of a problem and how it was solved. There are no "right answers" to these cases. The cases in this book ask **YOU** to:

- ***identify*** and ***evaluate*** real-life fisheries conservation and management problems;
- ***collaborate*** with colleagues to design creative approaches to solve the complex natural resource problems represented by these cases;

- ***evaluate*** existing information, ***identify*** and ***apply*** pertinent information, evaluate critical gaps in information, and design strategies to fill those gaps;
- ***create*** a range of possible solutions to complex problems;
- ***evaluate*** the pros and cons of various solutions from a range of competing stakeholder perspectives; and
- ***communicate*** the results of your problem-solving efforts in language that is understandable to a range of audiences.

So, this is not a typical textbook that is filled with information. Rather it is a series of thinking challenges that are designed to help you develop crucial professional skills. We (the authors) have no idea what information you will need to develop the approaches and solutions that you might propose, so it is impossible for us to collate it all here. But, even if we could, that information would quickly become outdated. As a professional you will need to develop effective strategies to seek out and evaluate current and valid information yourself, and then determine how you will use that information in your own problem solving. We cannot do that for you, just as nobody will be able to do it for you when you are faced with challenges as a practicing professional in the future.

These case studies offer you the opportunity to engage in exactly the types of thinking that effective professionals employ. Leave behind your old approach of relying on your instructors to tell you what you need to know and what to do with it. They will soon become part of your past, and you will be facing challenges like those described in this book without them being there to guide your every step. When you begin interviewing for jobs with natural resource agencies, non-governmental organizations, or other employers, you can expect many of them to employ a "What if . . ." line of questioning that will look very similar to the challenges in these cases. Students who willingly embrace the thinking challenges that we present here, and use these exercises as an opportunity to sharpen their thinking skills and develop new skills, will be the most successful in these interviews and, ultimately, as practicing professionals.

What other skills will employers expect you to have? Communication, teamwork, critical-thinking, and computer/information-literacy skills are among the top-desired skills for potential employees most frequently cited in surveys of employers (NACE 2010). I've already mentioned the importance of the latter two in that list. The cases in this book will also offer you the opportunity to develop the first two skills listed—communication and teamwork. Your instructor will likely ask you to work in teams on some of the cases in this book. Many students say that they dislike team projects, but filling a critical role within a professional team is exactly the way that you will be operating during your professional career. Take these cases as a new opportunity to learn effective teamwork: identify a common goal, draw on team members' strengths, deal with the inevitable personal conflicts that arise, and complete a task that might have been too large or complex for a single person. Finally, you will be asked to do a substantial amount of both oral and written communication related to these case studies. Effective communication is the skill listed as *most important* for successful applicants by many potential employers, and it is the area in which we (university faculty) hear the most complaints from employers about poor preparation in students. Take these cases as an opportunity to work on both your oral and written communication skills, and you will increase both your attractiveness to employers and your ultimate effectiveness as a practicing professional. I have had students use their writing assignments related to these cases as examples of their professional writing capabilities when they apply for jobs (or graduate school), and employers have been impressed when these writing samples demonstrate clarity, conciseness, and good critical and creative thinking in attacking a natural resource problem.

So, in closing, I ask that you think of the challenges presented by these case studies as an opportunity for you to practice and improve the very skills that you will employ in your natural resources career. Your course that uses this book will be different than most courses you have taken. Put aside your old thoughts about what a course should look like and how it should be conducted. You are going to work harder with these cases than you have in most courses in your student career. But, embrace this opportunity to do something significant with information, rather than just accumulating it. If you do this, you will be building skills that will help you both get a job and make valuable professional contributions throughout your career.

Brian R. Murphy
Virginia Tech
29 June 2010

References

NACE (National Association of Colleges and Employers). 2010. Job outlook 2010. NACE, Bethlehem, PA. Available: http://www.naceweb.org/SearchResult.aspx?keyword=job+outlook+2010 (July 2010).

Case 1

A Tale of Two Oceans: The Demise of Bluefin Tuna

In the Northern Atlantic Ocean, Off Greenland

The bluefin tuna saw the pod of false killer whales slash through the tuna school again, injuring several fish. The school pressed on at high speed through the chilly northern Atlantic, leaving both the victims and attacking predators behind. This tuna was built for speed. Its cylindrical body and sickle-like tail are shaped for maximum hydrodynamic efficiency. Even its dorsal fin can retract into a groove on the side of the body, thereby reducing high-speed drag as it pursues prey.

Bluefin tuna have been clocked at over 55 mph (90 km/hr). In fact, the English word 'tuna' is derived from the Greek verb 'thuno,' or 'to rush,' in reference to their hurried lifestyle. All tuna species are constantly on the go, needing rapid forward motion to force enough water over their gills to satisfy their body's tremendous oxygen demand. To satisfy this demand, tunas have larger gill-surface areas than most fish in order to extract as much oxygen as possible from the surrounding seawater. Additionally, the blood of tunas contains the oxygen-binding compound myoglobin, as well as the hemoglobin found in most animals. Myoglobin can carry even more oxygen than hemoglobin, thus providing additional oxygen to the tuna's heavily used swimming muscles. Finally, the muscle tissue of tunas contains more blood vessels than the muscle of other fishes, making the meat of tuna much darker in color than most other species found in fish markets worldwide.

Heat is also an important consideration for these marine sprinters. Unlike other fishes, which are generally considered to be 'cold-blooded,' many tuna species actually generate heat in their muscles through constant exertion and they display metabolic rates that are almost as high as those of mammals. Tuna have an elaborate countercurrent blood-flow system (the 'rete mirabile') that allows them to capture this heat and recycle it to swimming muscles, thereby increasing the muscle's power output even further.

At 335 lbs. (152 kg), this bluefin tuna was considerably larger than all of the other tuna in the school. At one time 'giant' bluefin (those over 310 lbs. or 141 kg) were common in both the northern and southern oceans, but heavy fishing effort in recent decades has increased harvest and decreased the aver-

FIGURE 1.1. Bluefin Tuna (*Thunnus thynnus*)

age size of fish. Bluefin once reached 40 years of age, 10 ft (3.1 m) in length and 1,500 lb (679 kg) in weight. Now, fishermen are lucky to catch fish that run 50–150 lbs (23–68 kg), which generally are less than 10 years of age.

This tuna was carrying a large battery-powered radio tag, attached to the muscle on its back near the dorsal fin. The tag had been placed there some 9 months earlier by scientists with the National Marine Fisheries Service (NMFS) in the United States (USA). The tuna had been caught by a recreational fisherman in the Gulf Stream, off the central coast of the eastern USA. It had taken two hours for the angler to bring the giant tuna to the boat. The U.S. recreational fishery for bluefin is highly regulated, and boats are limited to just a single fish that can be landed each year. This boat had already landed a fish a few months earlier, so our tuna was simply brought to boat-side and the wire leader was cut to release it. Before release, however, the radio tag (known as a pop-up satellite tag) was embedded in the dorsal musculature of the fish with a dart tag. The fish swam away, carrying the satellite tag that immediately began recording critical information about the fish: water depths visited, surrounding water temperatures and light levels, and the geographic location of the fish. These data were periodically logged in the memory of the tag, creating a permanent record of the environment and travels of our fish.

The Economics of Fishing

Fisheries for Highly Migratory Species (HMS) are divided between commercial and recreational fishers, each having different objectives and using different gears. Within the European Union (EU), HMS fisheries are dominated by commercial interests, while the recreational fisheries are much smaller. Conversely, in the USA, the economics of the recreational fishery dwarf commercial value. The output of all commercial fisheries in the USA was valued at approximately $1.5 billion (USD) in 1999 (NMFS 2001); by comparison, the 2001 total economic value of marine recreational fisheries in the USA was $8.3 billion (USD) (USDI et al. 2002). Similarly, the economic value of the recreational segment of USA fisheries for HMS dwarfs that of the corresponding commercial fishery (in contrast to the European fishery where the situation is reversed). In 2003, the NMFS issued over 22,000 permits for recreational fishing for HMS, as compared to less than 300 commercial longline permits for the same species (Moore 2003).

The entire landed value of all commercial HMS fisheries in the USA is only $120 million (USD) annually, with bluefin landings amounting to only $19 million (USD) of that annual total in 1999 (NMFS 2001). Recreational fisheries, including all their expenditures and associated employment, are economically critical to many coastal communities. For example, a single coastal city in the USA (Ocean City, Maryland) receives more than $20 million (USD) in economic benefits from a single week-long tournament for HMS (Moore 2003); and such tournaments and other charter boat enterprises operate along the entire continental coast.

Crossing the Atlantic

The large bluefin broke from the school of smaller tuna that it had traveled with for several days. While young tuna are gregarious schoolers, larger tuna are generally more solitary. This tuna had been moving constantly since she left the Mediterranean spawning grounds. In four months she had traveled over 4,800 miles (7,700 km) to near Newfoundland on the North American coast. Such a journey is not unusual for this species. A tagged bluefin released southeast of Japan was recaptured off the coast of the Mexican state of Baja California, after a migration of almost 6,700 miles (10,800 km) (Joseph et al. 1988). Another tagged bluefin traveled from the Gulf of Mexico to Norway in just 19 days; as a minimum straight line distance, this equates to swimming more than the length of its body every second for 19 days.

Each of the past 20 years our fish had returned to her birth area near Sicily, where large aggregations of bluefin tuna gathered to spawn. However, bluefins seemed to be less abundant there each year. Where once there had been multitudes of giant tuna in the aggregations, now the fish were far smaller and definitely fewer in number. Our giant had narrowly avoided death while in the Mediterranean this year. When she had entered the narrow inlet near Favignana, she found her way blocked by a wall of netting. Abundant boat traffic above had spooked the giant tuna, and she turned and fled the area. Unable to find the familiar spawning aggregations, she had headed back toward open water and eventually the open sea of the Atlantic.

The Mediterranean Sea, Near Sicily

Carlo Fucarini saw the giant tuna just outside the net, and he shouted to his fishing companions. "We've missed the largest fish! Look! It is the size that our fathers caught!" But it was too late. As they started hoisting the net, the giant tuna turned and disappeared into the deeper water.

As in coastal communities from Gibraltar to Italy, the men of Favignana have been catching bluefin tuna for longer than anyone can remember. Aristotle described Mediterranean tuna trapping in 350 BC, and Homer mentioned it in his classic, 'The Odyssey.' The mattanza (literally, 'the slaughter') is a community effort to harvest the bounty of large bluefin tuna that returned to the Mediterranean each spring to spawn (Gangi 2004). Large tonnara (trap nets) are placed along the coast, and migrating tuna are steered into large cages by the long wings of the net. The fishermen then herd the fish into a small cage and lift the net to concentrate the fish for dipping, spearing, and gaffing. This bloody harvest traditionally captured only the largest fish, which tended to lead the migratory schools. Many of the captured fish had already spawned, and medium and small fish were left to replenish the population,

FIGURE 1.2. The mattanza is a community effort to harvest the large bluefin tuna that return to the Mediterranean each spring.

so the mattanza has been a sustainable harvest for centuries. However, over recent decades both the size and number of fish captured has declined precipitously (Maggio 2000). In 2003 and 2004, the tonnara at Favignana failed to capture *any* fish (Rosenblum 2004), to the intense disappointment of both the fishermen and the tourists gathered to watch this ancient spectacle. The locals blame the increasing factory fishing that captures tuna in the open ocean before they have a chance to spawn in the Mediterranean. Where once there were over 200 tonnaras along the Mediterranean, now even the few remaining can no longer be supported by the meager catch.

Open Access in International Fisheries

Most fisheries in international waters (i.e., outside the exclusive economic zones) have been viewed historically as 'open access' (Ostrum 2000). This means that there are no limits on who is authorized to use the resource, and the fisheries can be freely exploited by anyone who chooses to invest the effort. At times throughout history and in various places, some groups (families, tribes, clans, and even nations) have laid claim to specific fisheries and defended local fisheries production as theirs alone (Nielsen 1976). However, even within these groups, the fisheries tend to be viewed as open access with no limitation beyond group membership to gain access to the fishery. Nobody limits the harvest in a true open-access regime, and resource depletion is the inevitable result (Ostrum 2000).

To protect what they consider being "their" fisheries resources, all coastal nations since the 1970s have claimed EEZs (exclusive economic zones) that extend 200 miles from their shores. All foreign fishers are excluded from

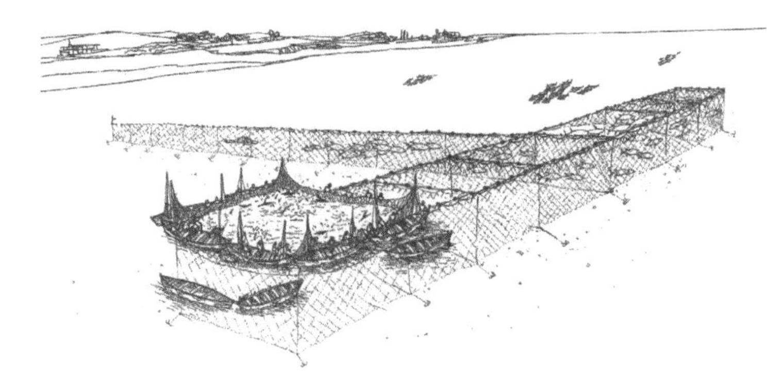

FIGURE 1.3. A tonnara (trap net), used to capture migrating tuna in the Mediterranean. Migrating fish are intercepted by the net wing set perpendicular to shore, and the fish are channeled into net cages from which they are dipped.

these zones, unless they have specific permits from the host government. These territorial claims have led to some 36% of the oceans' area to now be claimed by individual nations; these EEZ's include some 90% of the exploitable commercial fisheries stocks in the world. Nations manage their territorial fisheries with varying levels of effectiveness in terms of protecting fish stocks, but some species are captured by fishermen operating outside the EEZ of any nation. Who works to prevent common-property overharvest of these species?

The bluefin tuna and a series of other highly migratory predator species fall in the subgroup of fisheries that is heavily exploited in "international waters" (those outside of the EEZs). These highly migratory species (HMS) crisscross the globe (through both international waters and EEZs) in the course of their normal life cycle. Yet, by design, they are conspicuously absent from most nationalistic fisheries claims (some South American nations still claim control over tuna in their EEZs). Instead, similar to the highly migratory whales, these species are managed through various international agreements that control the harvest primarily through restrictions on the sizes and quantities of tunas landed in the various market countries.

Atlantic bluefin tuna management is overseen by an international treaty organization, the International Commission for the Conservation of Atlantic Tunas (ICCAT). The commission is made up of some 20 member countries on both sides of the Atlantic, plus Japan (which is a major fisher, importer, and consumer of Atlantic tunas). ICCAT's charter specifies that tuna stocks be managed for maximum sustainable yield (MSY), which is a management approach that seeks to maximize the available annual harvest from the fish stock by holding the population at roughly 50% of the normal environmental carrying capacity. At this abundance the population reproduces at its maximum rate, so the recruitment (addition) of new fish to the population is maximized each year. As long as the population is not reduced below this MSY level, fishermen can harvest the MSY (annual surplus production) indefinitely (Shaefer 1954).

Many scientists argue that MSY is not a truly achievable goal (Larkin 1977; Mace 2001), due to the inherent statistical uncertainty of the population measures developed to justify various harvest levels and due to the random nature of climate and other natural events that may cause wide swings in the natural abundance of fish. Bluefin tuna populations rise and fall with known and unknown changes in the environment. Biologists' ability to detect these changes, through limited statistical sampling that has its own inherent level of uncertainly, leads to fish stock estimates whose statistical confidence intervals may span several-fold or even an order of magnitude difference. Coupling this statistical and environmental uncertainty with the inherent wish of all fishermen to maximize catch as a return on their investment, management commissions such as ICCAT are under constant pressure from various economic interests to maximize the total allowable catch (TAC) under MSY by utilizing the most optimistic estimates of various population statistics. Estimating the size of fish stocks and recommending catch ceilings that will insure protection of the stock while allowing fishers a reasonable return on their investments is not an exact science or an easy job for any species, much less the highly migratory bluefin.

"Pity the fisheries biologist. Their job is to estimate the number of fish in a stock that should be safe to catch. They can't see the fish, yet they are somehow expected to figure out how many there are, what sizes they are and how many can be caught without diminishing next year's catch. Fisheries biologists work with mathematical models based on assumptions that they know could be wrong, for many features, such as growth, reproduction, and mortality rates, even in species that have been fished for centuries, may be poorly known. Fisheries biologists depend on counting techniques that they know are anything but perfect and rely on information from fishers reporting their catch, knowing that they may be lying, or misreporting, as the euphemism goes. Fisheries biologists know that the regulations proposed to enforce their estimates will be broken in most imaginative ways by the fishers who believe them to be intrusive. And they are all too aware of the impact of unexpected, unpredictable environmental events, such as shifting sea currents or the onset of an El Niño that can radically affect reproductive success and the future size of stocks in ways that can only be guessed at.

To make their estimates, they [fisheries biologists] *need to know as much as they can about the life history of the fish concerned... When fisheries biologists can't answer these questions* [of basic life history], *and that can happen often, they have to guess and then guess at how accurate their guess is likely to be.*

As scientists, as ecologists, fisheries biologists have long since learned to live with uncertainly and to accept the fact that their best estimate of stock size even at the best of times is still likely to be within 30 percent, plus or minus, of the real stock size. Through long experience they know that both fishers and politicians have little patience with such uncertainty... The biologists know, in fact, that whatever their estimates may be, the highest estimate of the stock size that they offer will be considered the one to go with. Moreover, managers and politicians may ignore the numbers because they are more concerned about the local employment of fishers. Virtually every politician knows that to take away someone's job is to also lose that person's vote."

(Berrill 1997; p. 40–43)

When it comes to protecting the spawning stocks of Atlantic bluefin tuna, ICCAT has not been immune from either the inherent statistical uncertainty of population estimates or the political pressures that cause science to be ignored in favor of short-term economic gain (Sloan 2003). ICCAT has failed to effectively protect the spawning stocks of Atlantic bluefin for many years, and population levels have fallen far below the level required to sustain MSY. Between 1970 and 1995, the recruitment (addition) of young fish into the adult population fell from over 300,000 per year to only 50,000 fish or fewer (Safina 1993; Chambers and Associates 2004). In 1975 stock size estimates for spawning fishes (those over 320 lbs, or 145 kg) were at only 25% of levels estimated in 1960; by 1990 spawning stock was at only 7% of the 1960 level (Ellis 2003).

One of the major disagreements within ICCAT revolves around the migratory patterns and resultant interbreeding and genetic stock structure of bluefin tuna populations in the northern Atlantic. ICCAT operates under an assumption that there are two distinct stocks of northern Atlantic bluefin: one that spawns in the Gulf of Mexico and migrates northerly along the coast of North America, and another stock that spawns in the Mediterranean and migrates along coastal Europe and northern Africa. The western stock seems to have declined precipitously over recent decades, and the catch from the western stock is highly regulated by ICAAT and the NMFS (National Marine Fisheries Service) of the United States. Conversely, the eastern stock has been relatively stable, and under generous catch limits. In the late 1990s, NMFS scientists began to suspect that western-stock fish were actually crossing the Atlantic into the eastern ICAAT regulatory zone, where they were being exploited by fishers not under the strict catch controls required to protect western fish. Tagging studies confirmed that some fish hatched in the Gulf of Mexico found their way to the Mediterranean as sub-adults or adults, but tag returns were very limited: eastern fishermen tend not to report the capture of tagged western fish in the Mediterranean because they fear it would cause the imposition of restrictive harvest quotas. Still today, European ICCAT representatives are reticent to redefine bluefin stock structure in order to avoid having to consider restrictive quotas on eastern fishermen. The NMFS has started new 'fishery-independent' research programs on bluefin migration;

these programs do not rely on fishermen to report the capture of tagged fish in order for migratory movements to be recorded (Lutcavage et al. 2001).

Tuna Farming

One of the major causes of the recent accelerated decline of bluefin is a practice known as 'tuna farming.' Small tuna are captured in the open ocean (or the Mediterranean) by purse seining, a practice that involves towing a large net (many miles or km long) around a school of tuna that has been sighted from airplanes or with sonar. The school is completely surrounded with a wall of net, and the bottom of the net is drawn in, like a purse string, to create a large bowl of net with the fish in it. The bowl is then winched to the boat, where tuna can be dip-netted or gaffed into the boat. However, in seining for tuna farming the fish are never actually landed by the seiner. The ship slowly tows the captured school of tuna to a quiet inshore area, where the fish are turned into large net pens where they will be fattened for months by feeding them other live fishes captured for this purpose. When the tuna have reached a size deemed large enough to go to market, the fish are harvested, iced, and shipped primarily to Japan for the sushi and sashimi markets. The Japanese market pays exorbitant prices for these bluefin, particularly fishes that have high fat content due to the time they spend in rich, cold northern waters.

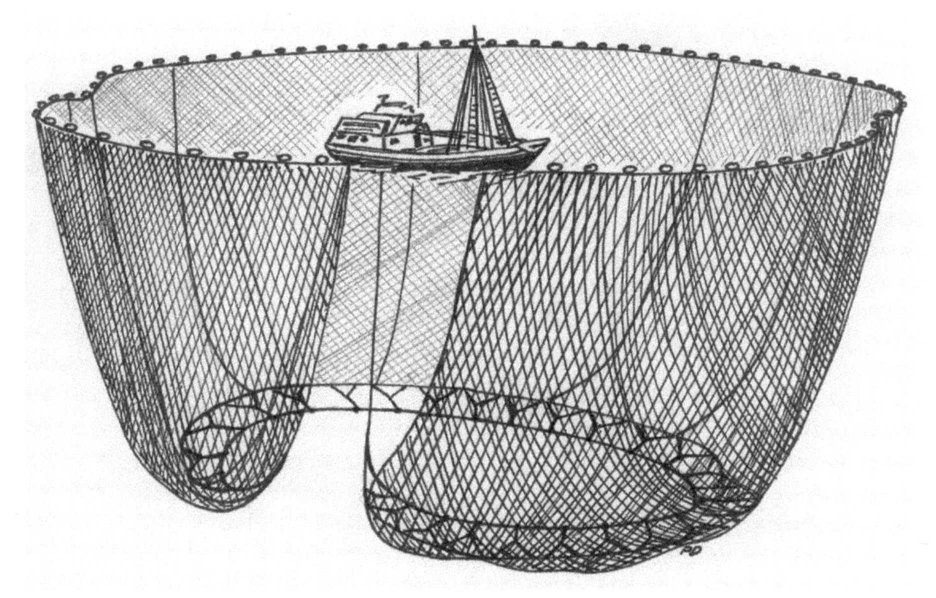

FIGURE 1.4. Illustration of a purse seine.

Now, here is the trick to tuna farming: when the fish are harvested from the net pens, they are classified as aquaculture products rather than wild fish (even though they originated from the wild). Tuna produced in aquaculture do not fall under the 'harvest' limitations of ICCAT, so the purse seiners and tuna farmers have found a loophole that leaves their industry completely unregulated. Purse seiners are capturing smaller tuna (including those not yet old enough to spawn) and more tuna than would ever be allowed by the ICCAT nations, and the final 'farmed' product also avoids international control. To add insult to injury, the seining and farming operations commonly kill up to 50% of the tuna handled, and they go to waste. Towing these very sensitive fishes in nets over long distances can also result in high mortality (Lato 2003), but the profit from even the limited number that are successfully raised and harvested still brings the operators a tidy profit. Tuna farming in the past decade has led to even further declines in juvenile bluefin of both the North Atlantic stock (through activities in the Mediterranean) and the Pacific stock of southern bluefin tuna *Thunnus maccoyii* around Australia and New Zealand.

The State of the Species

In the USA in the early 1990s, the National Audubon Society and the World Wildlife Fund failed in lobbying efforts to list Atlantic bluefin tuna under CITES (Convention on International Trade in Endangered Species of Wild Fauna and Flora), which would have banned exports from the USA and required monitoring of international trade. However, both the Atlantic bluefin tuna and the southern bluefin tuna were added to the IUCN (International Union for the Conservation of Nature) Red List of Threatened Species in 2003, along with their relatives the albacore tuna and the bigeye tuna (IUCN 2004). Yet, due to international political pressure, ICCAT continues to allow the harvest of Atlantic bluefin tuna at levels that endanger the species even further. By 2004, only two significant year classes remained in the Atlantic bluefin tuna stocks: the 1993 year class and the 1997. If these are not protected and allowed to spawn, the very existence of the bluefin tuna in the Atlantic Ocean is threatened.

The Western Atlantic Ocean, Near Newfoundland

The bluefin had followed the Gulfstream Current north, joining and then leaving several schools of smaller tuna. The archival tag on its back intermittently recorded the depth at which it swam, and the temperature of the water. One year after the tuna had been tagged, an internal computerized calendar noted the anniversary. A weak electric current was automatically passed through the thin wire that attached the electronic module to the dart tag embedded in the fish. That current reacted quickly with the surrounding seawater, and in a matter of days the wire corroded through. Thus released from the dart tag, the buoyant data module popped to the surface of the ocean and began transmitting its stored data to a satellite overhead. Later, NMFS scientists would download the data from the satellite, and the cross-ocean travels of the large tuna would become part of the growing database on migratory movements of the species across the entire Atlantic Ocean.

The bluefin pressed onward through the chilly waters. It fed on herring, mackerel, and squid. The combination of the rich food and cold temperatures stimulated the tuna's body to store more energy reserves as fat, particularly in the belly area of the fish. The weight of the fish had now increased to some 444 lbs (201 kg). It had become a fish that would be highly prized by any commercial fisherman.

The bluefin spotted a mixed group of fishes feeding on an aggregation of squid. Each of the squid was easily visible despite the fact that it was night, as they glowed with a strange luminescence. The big bluefin engulfed the first squid it reached, and moved toward another. But something slowed its forward progress. The tuna increased its swimming effort, but its body only turned sideways, pulled by a strange force on its mouth. A strong line stretched from its mouth toward the surface, some 50 meters above. Each time that the tuna tried to turn, the force eased slightly but then increased again. It was trapped, and could not get the forward momentum necessary to continually supply oxygen-rich water to its gills. Without that movement, its gills' oxygen extraction efficiency was greatly reduced. Its body functions began to slow as it rapidly used the oxygen in its bloodstream. As brain function declined, the tuna dropped into a state of near-death torpor.

The Spanish longliner crashed through the waves of the North Atlantic off Newfoundland, as it had been doing for weeks. Each of the crew paused in their work to grip the nearest solid object, as the ship rolled violently in the large waves. These conditions made it obvious why commercial fishing in the North Atlantic is one of the most dangerous occupations in the world. It is 0100 hours, and they still have several miles of line left to set. At 40 miles in total length, the line has almost 3000 dropper lines. Each dropper must be baited by hand with a whole squid and a chemical light stick to attract the target species for this trip, the Atlantic swordfish.

0500 hours: After only 2 hours of sleep, the longline fishermen are back at their posts on deck. They will spend most of the day retrieving the line and processing their catch. By 0900 their catch amounted to only 18 swordfish, mostly small, and a few sharks and seabirds that they threw back into the sea. They won't pay for their trip at this rate. But suddenly the line went tight as the winch strained to lift it. The men saw a large fish rising up from the depth, and they shouted with excitement as they realized that it was a giant bluefin tuna. They hardly noticed the colorful dart tag on the back of the tuna, as they happily discussed their luck. This single fish will go a long way toward making their trip a financial success. It is obvious that the tuna is barely alive. They dispatch it quickly with a club blow to the head, unhook it, and route it to the hold below. There it is rapidly gutted and packed in ice. This fish is too valuable to let it spoil.

Longline Fishing

Our tuna was caught incidentally by a swordfish boat, but its fate was similar to many of its relatives. Most fisheries for bluefin tuna utilize longlining techniques. Longliners stretch up to 60 miles of floating lines across the ocean surface, with thousands of baited dropper lines to hook fish. But fish are not all they hook. Longline fishing is extremely controversial because of the unintended catch of many 'incidental' species: sharks, marine turtles, marine mammals, and even oceanic birds like petrels and albatrosses. Longliners set hundreds of millions of hooks each year throughout the world's oceans. The BBC reported that a loggerhead or leatherback turtle has a 40–60% chance of encountering a longline hook each year (Amos 2004). Over 60,000 albatrosses and petrels are drowned annually on longlines in New Zealand waters alone (RFBPS 2004).

Longline fishing can be lucrative for a few boat owners and captains, but generally the income is not far beyond the expenses for a fishing trip. Boats are expensive, the long distance to fishing grounds can eat up a lot of fuel, and costs are high for maintaining a crew on a trip that may last a few weeks to months.

Box 1.1. Tsukiji Fish Market, Tokyo

January 5, 2001 (AP). An enormous bluefin tuna—a fish prized as sushi—sold for a record $173,600 (USD) Friday in the first auction of the year at Tokyo's main fish market. At $391 per pound, the 444-pound fish was the most expensive auctioned off at the Tsukiji Central Fish market in years. In 1996, a 250-pound bluefin fetched $44,100. Called honmaguro in Japanese, bluefin tuna is popularly served raw as sashimi or sushi in restaurants where a plate of slices can command a bill of more than $100.

The economics of longline fishing for swordfish were detailed by a U.S. swordboat captain (Greenlaw 1999) associated with the fishery described in the book and film, *The Perfect Storm* (Junger 1997). A 4-week trip that landed some 51,000 lbs (23,087 kg) of swordfish cost over $51,000 (USD) in boat and crew expenses; bait alone cost almost $8,000 (USD). The trip

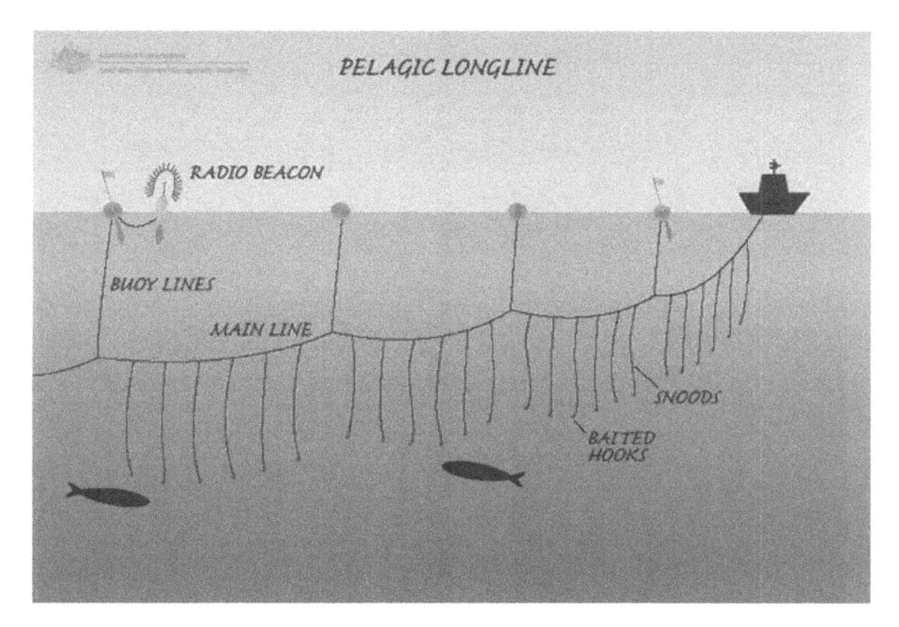

FIGURE 1.5. Diagrammatic representation of longline fishing.

produced over $41,000 (USD) in income for the boat owner, but only $5,484 (before taxes) for each crewmember. And this was a very good fishing trip; most trips do not produce nearly that much income for the deck crew, and the fishing season may last only a few months. Commercial longlining clearly is not a lucrative business for anyone except the most successful boat captains and their boat owners. One trip may be highly successful, while the next may not even pay the boat's expenses. Common property economics always lead to negligible net profit within the fishery as a whole; a few highly successful operators may reap large profits, but the average operator and fishermen barely break even and many go bankrupt. And the economics get worse as fisheries stocks are overexploited and fish populations decline. As a Japanese longliner captain expressed about his own industry:

"The total number of fish has been decreasing a lot. So the biggest problem for me is that my income has been reduced. I suspect we probably have no future if we keep doing this type of reckless fishing. In order to rebuild the tuna population, we have to control not only long-line fishing boats like ours, but also the net-fishing boats, since they catch even the small fish. Otherwise, I think the future of fishing is doomed."

Takeharu Jinguji, retired Japanese longline fisherman (PBS 2002)

Figure 1.6. Frozen tuna at the Tsukiji Fish Market.

Conclusion

'Commercial extinction' is a phenomenon that may have saved the great whales from disappearing in the middle of the 19th Century (WWF 2004); they became scarce enough that it was not economically viable to pursue them down to the last individuals. While whaling was a significant economic force in the 19th Century (and was continued for social reasons into the 20th Century), the fishery reached a point of diminishing returns, wherein it made economic sense to turn attention to other fisheries or alternative pursuits. The same phenomenon cannot be counted on to save the bluefin tuna. Individual tuna have an economic value that far exceeds that ever seen for the whales. Their extreme value for the elite sushi/sashimi market virtually guarantees that fishers will hunt down the last of the species, individual by individual (or at least to the point that the species may drift to extinction due to the lack of critical spawning aggregations). No fishery in the history of the world has ever produced such alluring profits as the bluefin tuna market, and that very value may be its undoing. Without a significant change in the attitudes of fishers toward conservation of the very resource that brings them wealth, or changes in the international control and management of bluefin fisheries (including 'farming'), the bluefin tuna is likely doomed to extinction. The sad fact is that the bluefin could be saved and the species could be managed as a sustainable resource indefinitely; but short-sighted entrepreneurs and their political power, coupled with international management ineffectiveness, will probably not allow that to happen (Clark 1974). Bluefin tuna won't be the first species that has disappeared for these reasons.

Discussion Questions

1. Are humans ethically bound to protect the existence of other species on the planet, or are we within our rights to use other species for our economic betterment, even if it means that they may be driven to extinction? Isn't extinction just a natural process that all species will face someday, one way or another?
2. How can the management of common property resources be improved? What impetus is there for resource users to cooperate with any management scheme?
3. Are bluefin tuna harvested by fishermen from your country or region of the world? How does your government regulate their catch?
4. Highly migratory species present complex conservation and management problems because they cross so many jurisdictional boundaries. How should nations cooperate to protect and manage such species? Is ICCAT's structure a reasonable approach to this problem? How could it be improved?
5. How should the decimation of juvenile bluefin stocks for tuna farming be prevented? Who has the authority (and the power) to protect these stocks? Should farmed fish products be regulated in the same way that wild capture fisheries are regulated? How can the high mortality of captive bluefin tuna be reduced? Should there be a penalty for ineffective husbandry practices, beyond the obvious loss of income, when the product dies?
6. The extreme value of bluefin tuna in the Japanese market is driving most tuna operations in the world today. So far, existing legal structures have been ineffective at protecting the bluefin. Can anything be done at the market level to curb the demand for this exorbitantly priced product? Should free-market economics be allowed to drive the overexploitation and possible extinction of a species? If you say no to this question, how can we prevent such from happening (particularly in an international market)?

7. Some scientists claim that, in the future, aquaculture will fill much of the demand for wild-caught fisheries products. What are the pros and cons of replacing wild-caught products with products from aquaculture? Do you think that large, predatory, pelagic species such as bluefin tuna can ever be successfully cultured in captivity?

References

Amos, J. 2004. Longlines imperil Pacific turtles. BBC (British Broadcasting Corporation), London, UK. Available: http://news.bbc.co.uk/1/hi/sci/tech/3485195.stm (December 2009).

Berrill, M. 1997. The plundered seas: can the world's fish be saved? Sierra Club Books, San Francisco, California.

Chambers and Associates. 2004. Giant bluefin tuna of the Atlantic—severity of the decline. Available: http://www.big,marinefish.com/bluefin.html (December 2009).

Clark, C. W. 1974. The economics of overexploitation. Science 181:630–634.

Ellis, R. 2003. The empty ocean. Island Press, Washington, D.C.

Gangi, R. 2004. Mattanza. Best of Sicily Magazine. Available: http://www.bestofsicily.com/mag/art125.htm (December 2009).

Greenlaw, L. 1999. The hungry ocean: a swordboat captain's journey. Hyperion, New York.

IUCN (International Union for the Conservation of Nature). 2004. The IUCN Red List of Threatened Species. Species Survival Commission. International Union for Conservation of Nature and Natural Resources, Cambridge, UK. Available: http://www.redlist.org/ (December 2009).

Joseph, J., W. Klawe, and P. Murphy. 1988. Tuna and billfish—fish without a country, 4th edition. Inter-American Tropical Tuna Commission, La Jolla, California

Junger, S. 1997. The perfect storm. W. W. Norton & Company, New York.

Larkin, P. A. 1977. An epitaph for the concept of maximum sustainable yield. Transactions of the American Fisheries Society 106:1–11.

Lato, D. 2003. $6 million tuna lost as cage collapses. GROWfish News. Available: http://www.growfish.com/au/Grow/Pages/News/2003/mar2003/56603.htm (December 2009).

Lutcavage, M. et al. 2001. Results of pop-up satellite tagging of Atlantic bluefin tuna. Microwave Telemetry, Inc. Available: http://www.microwavetelemetry.com/Fish_PTTs/bluefintuna.htm (December 2009).

Mace, P. M. 2001. A new role for MSY in single-species and ecosystem approaches to fisheries stock assessment and management. Fish and Fisheries 2:2–32.

Maggio, T. 2000. Mattanza: Love and death in the Sea of Sicily. Perseus, Michigan.

Moore, H. 2003. Testimony for oversight hearing on upcoming 18th regular meeting of ICCAT, before the House Subcommittee on Fisheries Conservation, Wildlife and Oceans, 28 October 2003. Recreational Fishing Alliance, New Gretna, New Jersey. Available: http://resourcescommittee.house.gov/108cong/fish/2003oct30/moore.htm (December 2009).

NMFS (National Marine Fisheries Service). 2001. Economics of HMS fisheries. SAFE Report. NOAA Fisheries, Silver Spring, Maryland. Available: www.nmfs.noaa.gov/sfa/hms/Safe_Report/Safe_5.PDF (December 2009).

Nielsen, L. A. 1976. The evolution of fisheries management philosophy. Marine Fisheries Review 38(12):15–23.

Ostrum, E. 2000. Private and common property rights. Pages 332–379 *in* B. Bouckaert and G. De Geest (editors). Encyclopedia of law and economics, Vol. II. Civil law and economics. Elger, Cheltenham, UK.

PBS (Public Broadcasting Service). 2002. Empty oceans, empty nets: the race to save marine fisheries. Available: http://www.pbs.org/emptyoceans/tuna/ (December 2009).

RFBPS (Royal Forest and Bird Protection Society of New Zealand). 2004. Marine and coastal campaigns: save the albatross. RFBPS. Wellington, New Zealand. Available: http://www.forestandbird.org.nz/marine/index.asp (December 2009).

Rosenblum, M. 2004. Bluefin tuna losing battle for survival. MSNBC News. Available: http://www.msnbc.msn.com/id/5428979/ (December 2009).

Safina, C. 1993. Bluefin tuna in the west Atlantic: Negligent management, and the making of an endangered species. Conservation Biology 7:229–234.

Schaefer, M. B. 1954. Some aspects of the dynamics of populations important to the management of commercial marine fisheries. Inter-American Tropical Tuna Commission Bulletin 1:25–56.

Sloan, S. 2003. Ocean bankruptcy: world fisheries on the brink of disaster. The Lyons Press, Guilford, Connecticut.

USDI (U.S. Department of the Interior, U.S. Fish and Wildlife Service, U.S. Department of Commerce, and U.S. Census Bureau). 2002. National survey of fishing, hunting ,and wildlife associated recreation. Washington, D.C. Available: www.census.gov/prod/2002pubs/FHW01.pdf (December 2009).

WWF (World Wildlife Fund). 2004. WWF South Pacific whale campaign. Available: http://www.wwfpacific.org.fj/whales_campaign_depleted.htm (December 2009).

NOTES

NOTES

Case 2

What's for Dinner? Environmentally Conscious Seafood Choices

> "Our enjoyment of these foods is heightened if we also know something of the creatures from which they are derived, how and where they live, how they are caught, their habits and migrations... Each of the millions of people who buy and eat fish can play an active part in conservation."
> —Rachel Carson, 1943

Although you're usually very opposed to going on blind dates, your friends have convinced you that Alex is worth making an exception for. The only thing your friends will tell you is that Alex is very environmentally and socially conscious. And that Alex loves seafood and people who know how to cook.

You obviously want to impress your date, but you are concerned about picking the right seafood. Taste is no problem—you are an excellent cook and can make even liver taste delicious. You have been hearing a lot in the media lately about how fisheries all over the world are in trouble, due to habitat degradation, overfishing, bycatch, and a number of other problems. You would hate to unintentionally contribute to environmental problems just by making dinner.

So which seafood should you choose? Some fisheries catch a lot of bycatch of other species, including marine mammals, birds, and sea turtles; some fish populations have been overharvested and have low abundance; some fish species are long-lived and slow to reproduce; some fish have consumption advisories from high levels of mercury and other contaminants; some types of seafood are imported; some seafood production causes habitat degradation or risk to other species through poor aquaculture or fishing practices. How can you choose?

The fishmonger at your local grocery store has the following five types of seafood available:

1. Atlantic bluefin tuna (*Thunnus thynnus*), caught in Spain
2. Orange roughy (*Hoplostethus atlanticus*), caught in New Zealand
3. Chilean sea bass (*Dissostichus eleginoides*), caught in Argentina
4. Atlantic cod (*Gadus morhua*), caught in United States
5. Farm-raised Atlantic salmon (*Salmo salar*), raised in the United States

Use the information on Sustainable Seafood at the Blue Ocean Institute's website (http://www.blueocean.org/seafood/seafood-guide) to learn more about these types of seafood.

As you look through the information on these fish species, think about what is personally important to you. Are you most concerned about impacts on other species through bycatch; impacts to habitat through harvest and aquaculture methods; sustainability of the fish population; buying locally or buying from less wealthy nations; consuming seafood that is free of contaminants; other factors related to the seafood industry?

FIGURE 2.1. Atlantic cod *Gadus morhua* and Atlantic salmon *Salmo salar*.

Assignment

Select one of these five fish species to serve at your dinner (you can fill in the tables below to help you keep track of your research). You know your date is going to ask you about your choice (or it's a good conversation starter during a lull). To be prepared to defend your seafood selection, write a 1-page justification of your choice that you can peek at during dinner. Just giving your date the scores from the Blue Ocean Institute website is insufficient. There is no right or wrong answer; your choice depends on what aspects you think are most important and why.

Discussion Questions

Check out your favorite seafood on the Blue Ocean Institute website—how does it rate? What factors impacted its rating? How will what you learned from this site affect your future choices of seafood when you cook or are at a restaurant?

1. What type(s) of seafood might you recommend that your fishmonger start selling, based on the information on this website?

2. Should this information be available at grocery stores and restaurants? If it was, how would that positively and negatively impact fish populations, fisheries and aquaculture, and the public? Do you think it would affect what types of seafood people eat?

References

Carson, R. 1943. Food from the sea: fish and shellfish of New England. U.S. Government Printing Office, Washington, D.C.

Table 2.1. A template for information collecting.

Wild-caught Species	Life history	Abundance	Habitat, Gear	Management	Bycatch	Notes
Atlantic bluefin tuna						
Orange roughy						
Chilean sea bass						
Atlantic cod						

Farm-raised Species	Oper. risks	Feed	Pollution	Risk to other spp.	Eco. effects	Notes
Farm-raised Atlantic salmon						

NOTES

NOTES

Case 3

A Float Trip on the Bad River, South Dakota: How Will Cyprinid Distributions Change from Headwaters to Mouth?

The Setting

The Bad River (Figure 3.1) flows through ruggedly beautiful country in the semi-arid region of western South Dakota. The watershed area for the Bad River is 8,044 km^2, and mean annual rainfall in the basin is 41–46 cm. For the years 1929–2000, average annual discharge was 4.9 m^3/sec, with annual means ranging from as low as 0.2 m^3/sec to as much as 34 m^3/sec (Milewski 2001). Your instructor will take you on a photographic "float trip" of the river, from its headwaters to its confluence with the Missouri River (a linear distance of approximately 120 km "as the crow flies").

Overview of this Exercise

As you travel from the headwaters to the mouth of the Bad River, consider both the changes in habitat as well as the influence of the Missouri River as you near the river mouth. You will be asked to predict expected patterns in abundance of two different minnows, the fathead minnow and flathead chub (Figure 3.2), from the headwaters to the mouth. We will provide a graph showing the changes in abundance for the two species. However, you will need to identify which trend line is most likely associated with each of the two species.

Initial Assignment

You will first need to research life history information for both the fathead minnow and flathead chub. What is their maximum size and maximum age? What type of habitat do they prefer? What are their food habits? What are their environmental tolerances for water quality, turbidity, water velocity, etc.? What are their interrelationships with predators? Sources of information can include both books at your library, or internet resources.

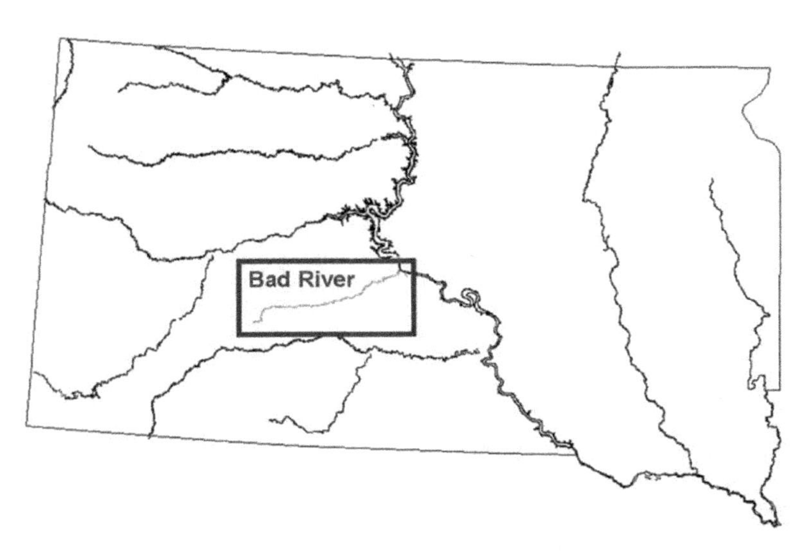

FIGURE 3.1. Map of South Dakota showing the Bad River, a western tributary of the Missouri River.

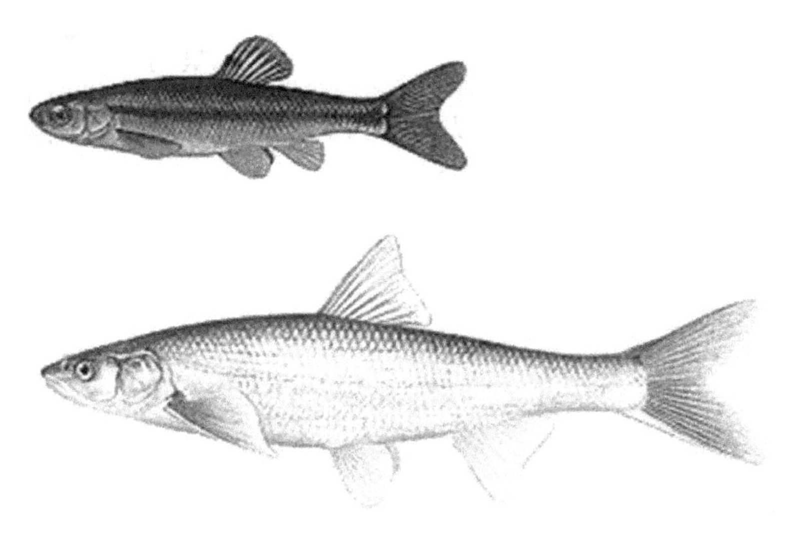

FIGURE 3.2. Fathead minnow (top) and flathead chub (bottom).

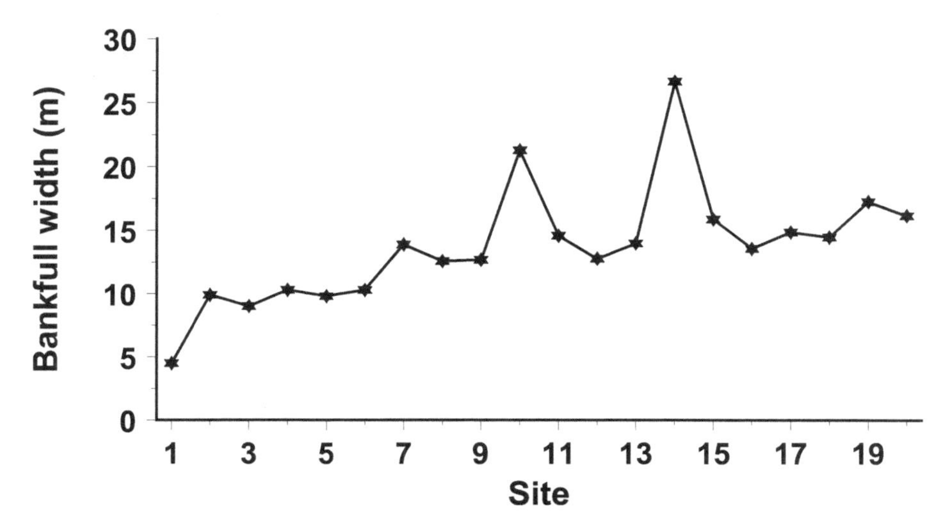

FIGURE 3.3. Changes in bankfull width (channel width between the tops of the most pronounced banks on either side of a stream reach) along the Bad River, South Dakota. Site 1 represents the headwaters and Site 20 is just prior to the confluence with the Missouri River.

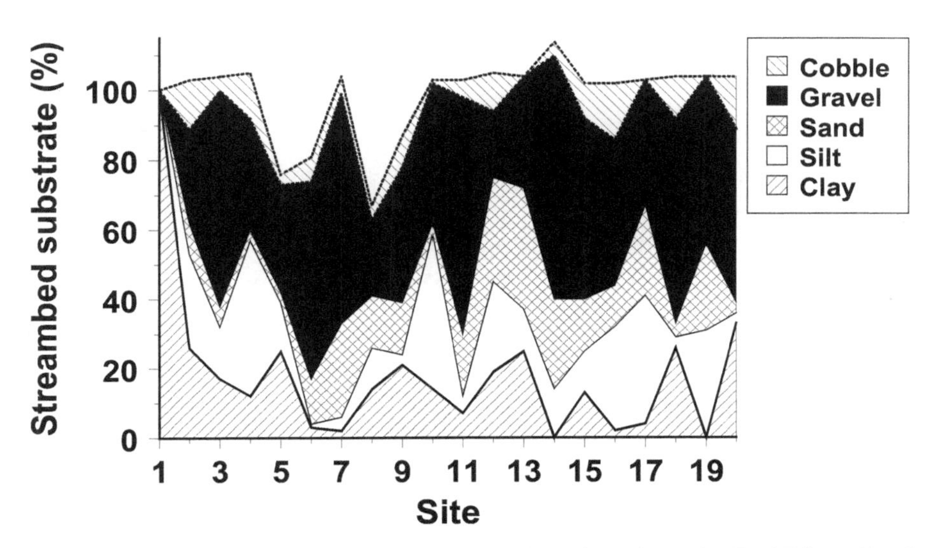

FIGURE 3.4. Streambed substrate (%) at sampling sites along the Bad River, South Dakota. Site 1 represents the headwaters and Site 20 is just prior to the confluence with the Missouri River. Percentages do not always add to 100 because of rounding error and other substrate types (e.g., boulders).

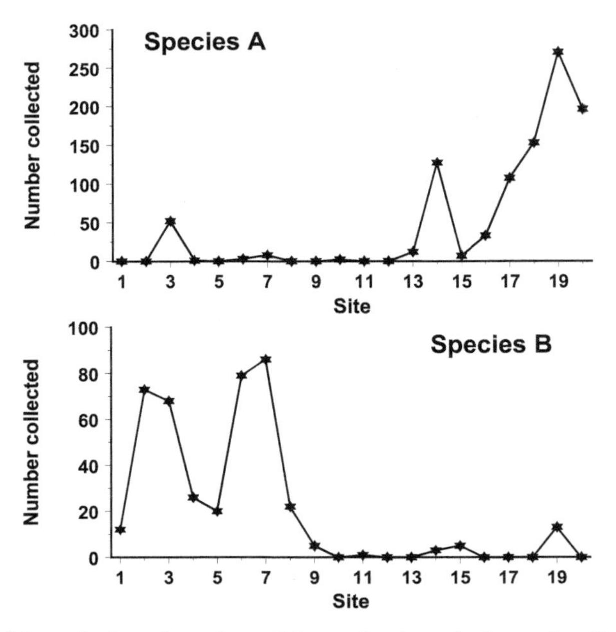

FIGURE 3.5. Numerical catches of two fish species, here designated as Species A and Species B, from the headwaters (Site 1) to the mouth (Site 20) of the Bad River, South Dakota. Fishes were collected with a bag seine (8-mm bar mesh) that reached from bank to bank, and pools, runs and riffles were seined in a downstream direction toward a block net.

Problem to be Solved

1. Consider Figure 3.3. Given the changes in channel width, what habitat characteristics might you expect to change from the headwaters to the mouth?
2. Consider Figure 3.4. Does substrate appear to change from station 1 through station 20? Are there any consistent trends in substrate types from headwaters to the mouth?
3. Consider Figure 3.5. Look specifically at the abundance of each species from headwaters to the Missouri River confluence.
4. Consider the life history information that you located for each species. Is the flathead chub Species A or Species B? Why? Is the fathead minnow Species A or Species B? Why?

References

Milewski, C. L. 2001. Local and systemic controls on fish and fish habitat in South Dakota rivers and streams: implications for management. Doctoral dissertation. South Dakota State University, Brookings.

NOTES

NOTES

Case 4

Effects of Angling on a Previously Unexploited Wisconsin Fish Community

Background

The effect of angling on fish population size structure is dependent on a myriad of factors, such as population abundance, the amount of angling effort, the size of the water body, and even differential vulnerability of different fish species. The opportunity to study unexploited populations can be extremely valuable, including situations where subsequent angler harvest is then assessed. Data sets for unexploited populations tend to be relatively rare. Information on unexploited populations can provide our profession with valuable examples of just what is possible when a population is not exploited. That way, realistic (i.e., feasible) management objectives might be set for various harvest regulations that a fishery biologist might select. In this case study, you will explore the effects of angling on a previously unexploited fish community in a small Wisconsin lake.

Setting

Mid Lake (4.7 ha) is one of a series of small lakes in Hartman Creek State Park, Wisconsin. The maximum lake depth was 1.8 m, and given the relatively clear water that allowed light penetration to the entire lake bottom, submergent vegetation growth was abundant throughout the lake. The fish community in Mid Lake included bluegills, largemouth bass, northern pike, pumpkinseeds, and yellow perch.

The lake was closed to all fishing from 1938 to 1976. On 1 May 1976, anglers were permitted to harvest any fish species, with seasons extending from early May through February. There were no length limits on harvested fish, and daily creel limits were liberal (five per day for largemouth bass, five per day for northern pike, and 50 per day for panfish). Random stratified creel surveys indicated angling effort of 230 hours/ha in 1976 and 62 hours/ha in 1979.

Hartman Creek State Park is located in central Wisconsin. Check their website for information on this scenic location (http://www.dnr.state.wi.us/org/land/parks/specific/hartman/index.html; site active August 2009).

Problem or Dilemma

Your instructor will break the class into small teams for discussion.

1. Assess the length-frequency histograms and age-structure data for bluegill, largemouth bass, pumpkinseed, and yellow perch collected by electrofishing from Mid Lake, Wisconsin prior to angler exploitation (Figures 4.1–4.4).
2. Predict the likely effects of angler exploitation on population size and age structure after angling has been allowed for one year.
3. What percent decline in population abundance would you predict for each species?
4. Why do you expect such effects? Be certain to consider the fact that different species may be differentially affected by exploitation. After presentation of your expectations to the class, your instructor will show you the changes that actually occurred.

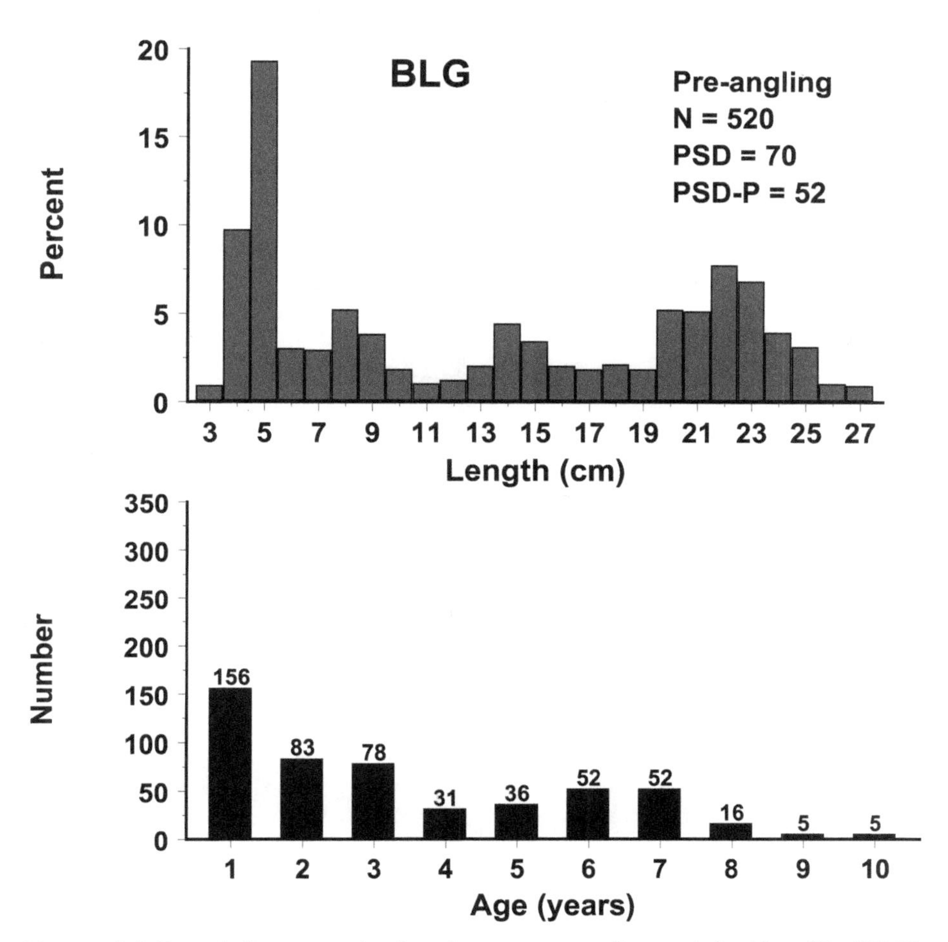

FIGURE 4.1. Length frequency (top) and age structure (bottom) for bluegills (BLG) collected from Mid Lake, Wisconsin by night electrofishing prior to angler exploitation. PSD = proportional size distribution (percent of 8-cm and longer bluegills that also exceed 15 cm)(Guy et al. 2007); PSD-P = proportional size distribution of preferred-length fish (percent of 8-cm and longer bluegills that also exceed 20 cm).

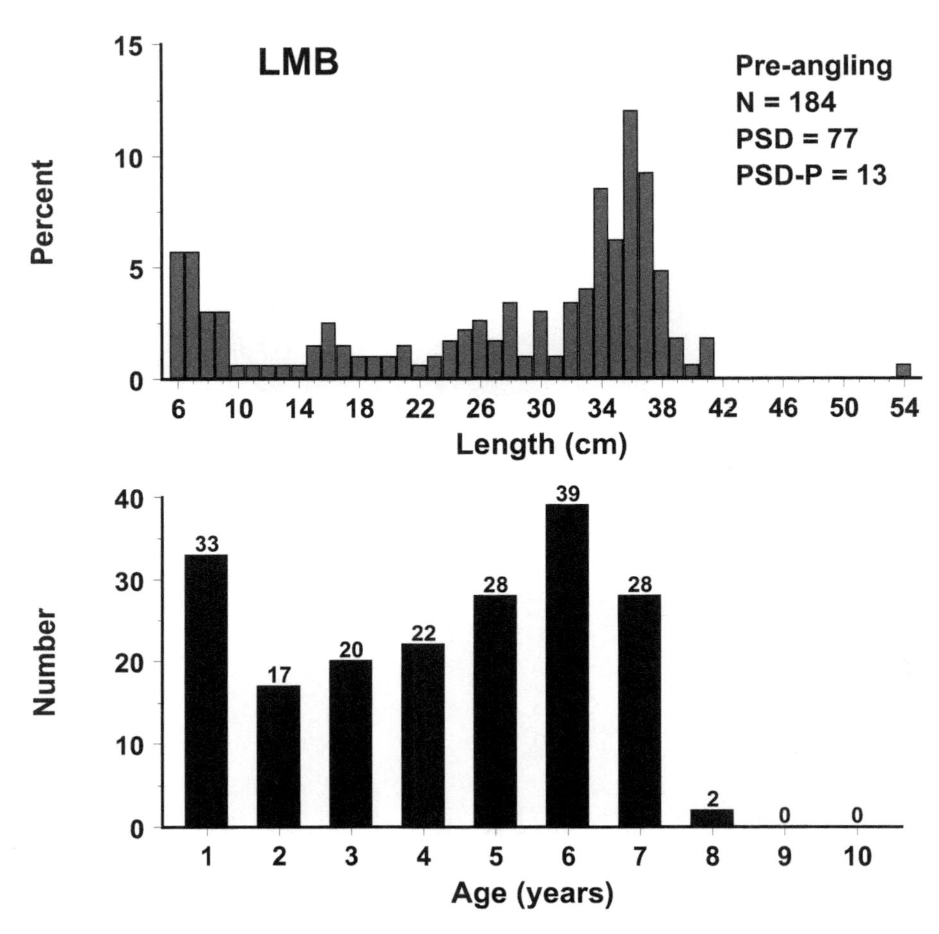

FIGURE **4.2.** Length frequency (top) and age structure (bottom) for largemouth bass (LMB) collected from Mid Lake, Wisconsin by night electrofishing prior to angler exploitation. PSD = proportional size distribution (percent of 20-cm and longer largemouth bass that also exceed 30 cm)(Guy et al. 2007); PSD-P = proportional size distribution of preferred-length fish (percent of 20-cm and longer bass that also exceed 38 cm).

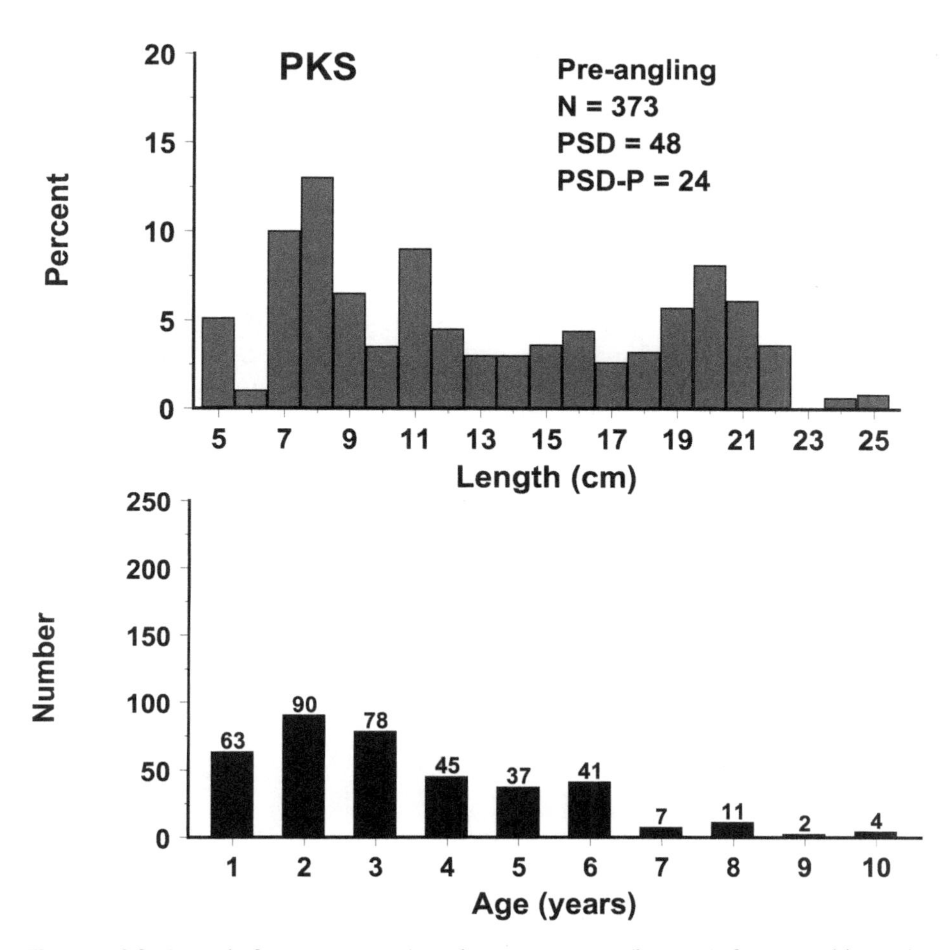

FIGURE 4.3. Length frequency (top) and age structure (bottom) for pumpkinseeds (PKS) collected from Mid Lake, Wisconsin by night electrofishing prior to angler exploitation. PSD = proportional size distribution (percent of 8-cm and longer pumpkinseeds that also exceed 15 cm)(Guy et al. 2007); PSD-P = proportional size distribution of preferred-length fish (percent of 8-cm and longer pumpkinseeds that also exceed 20 cm).

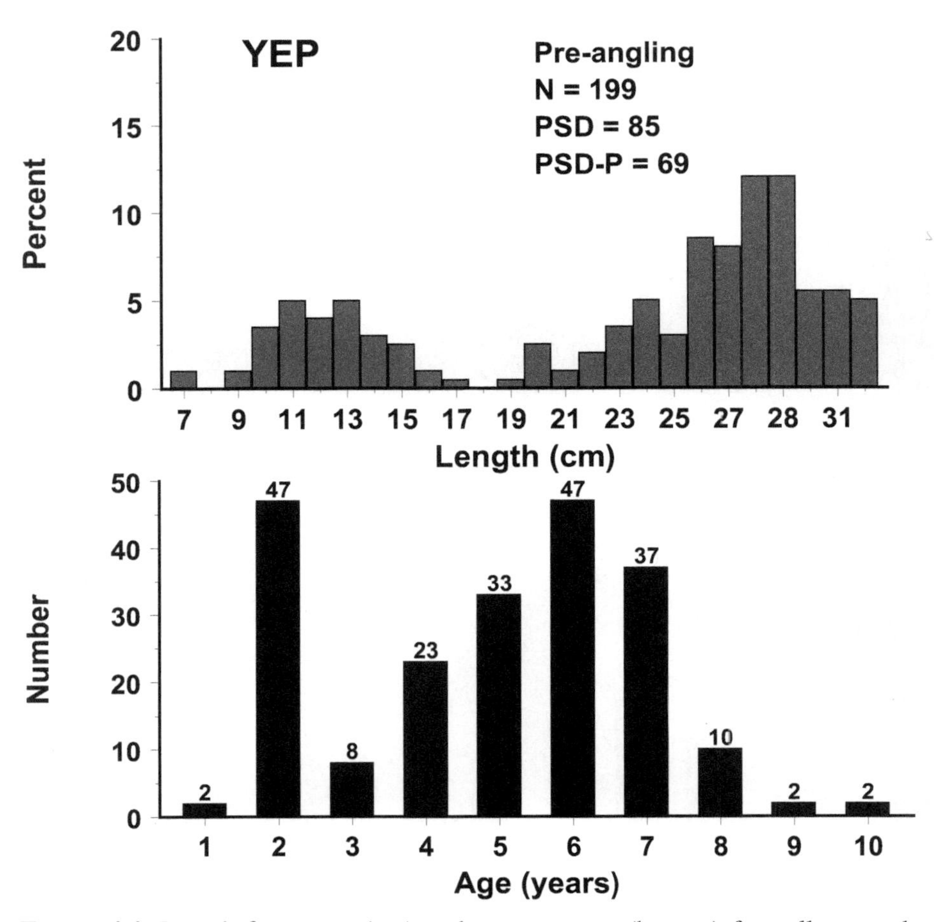

FIGURE 4.4. Length frequency (top) and age structure (bottom) for yellow perch (YEP) collected from Mid Lake, Wisconsin by night electrofishing prior to angler exploitation. PSD = proportional size distribution (percent of 13-cm and longer yellow perch that also exceed 20 cm)(Guy et al. 2007); PSD-P = proportional size distribution of preferred-length fish (percent of 13-cm and longer perch that also exceed 25 cm).

NOTES

NOTES

Case 5

Communism Meets the Tragedy of the Commons: A Fisheries Management Conflict in Rural China

The Problem

Presently, hundreds of commercial fishermen unrestrictedly harvest fish from Caohai Lake using trapnets, gillnets, and various small fish traps. The shallow lake is so choked with trapnets that it is difficult to cross the lake in a boat. Government officials would like to boost fisheries production in the lake in order to help provide food for the local populace. They know little of the current status of the fishery, and they fear local backlash to any management plan that they might implement.

The Setting

South-central China's Guizhou Province is home to some of the poorest people in that growing nation. Guizhou's Caohai Nature Reserve (CNR) is a national wildlife preserve established for the protection of the endangered Black-necked Cranes. The cranes breed in Siberia, and winter on Caohai ("Sea of Grass") Lake's 20 km^2 of wetland area. Human population is extremely dense in the 96-km^2 Caohai Lake watershed. Cranes were hunted almost to extinction before the International Crane Foundation (Baraboo, Wisconsin, USA) intervened in the early 1990s to help develop a protection and management plan that allowed cranes and farmers to co-exist. The heart of the plan was the establishment of a community micro-enterprise development grant program that helped the local people establish livelihoods that were not contrary to crane conservation objectives. Locals shifted to enterprises such as hog farming, native plant collection for the herbal industry, and floral horticulture. Fisheries on the lake also have been historically important to the region.

Water quality parameters for Caohai Lake indicate that the lake is almost hypereutrophic. Untreated sewage from all of the surrounding communities flows directly into the lake. Rooted macrophytes (emergent and submergent)

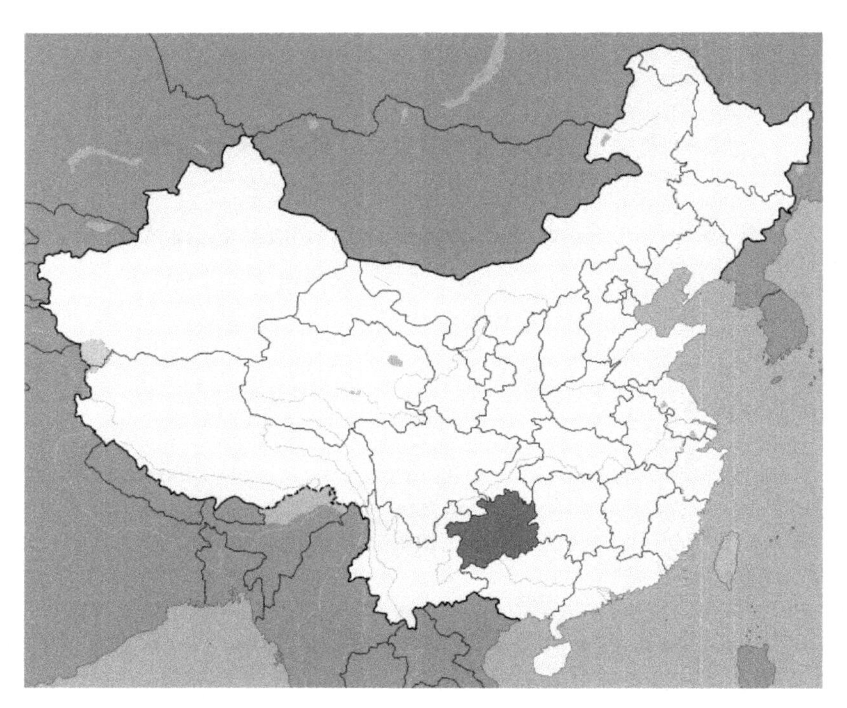

FIGURE 5.1. Dark shaded area indicates Guizhou province.

FIGURE 5.2. Caohai Lake is very shallow and heavily vegetated. Hand-cut channels are maintained to allow boat passage. There are no motorized boats found on the lake.

are abundant, which keeps the water clarity relatively high. The lake also receives runoff from numerous ore slag piles, which are the remnants of a once-thriving small-scale zinc smelter industry scattered throughout the watershed. The Weining County Planning Board currently is considering a number of options to reduce sewage runoff to the lake.

Natural resources on CNR are managed by a small refuge staff, which includes several wildlife biologists and resource economists. While cranes are directly protected on the refuge, there are virtually no restrictions on the exploitation of other natural resources. Wetland vegetation is routinely harvested by hand for basket-making and to feed livestock (primarily water buffalo). The fish populations of Caohai Lake have historically been an important protein source for the local populace. Analysis of fish sales in the local market shows that catch is dominated by eight species of cyprinids, the major component of which is crucian carp. Catch records indicate that commercial harvest from Caohai Lake peaked in 1989 at 600 mt; the latest catch records (2001) show an annual harvest of 250 mt. Total catch declined by 20% between 1999 and 2001. Governmental fisheries management efforts were abandoned in the early 1990s when irate farmers and commercial fishermen routinely attacked fisheries biologists and enforcement officers with farm implements. Published limits on trap net numbers and sizes are ignored, as are an annual licensing requirement and the closed season designed to protect spawning fishes.

The Mission

You were contacted by the local and federal management agencies to conduct a preliminary assessment of the present status of the Caohai Lake fishery, and to make recommendations for its improvement. You will be on-site at Caohai for four days. You have access to hand-poled boats operated by concessionaires on the lake to service tourists visiting the lake. You have a car and driver/interpreter at your disposal, but no fisheries sampling gear. The local federal fisheries biologists (*Bureau of Animal Husbandry and Fisheries, Weining County Office*) are available for consultation. Using the resources available, design a strategy to develop a set of management recommendations. Your recommendations will be presented to the Weining County Planning Board. This governmental board, which includes no scientists, is deeply involved with community well being, and its local politicians will be influential in supporting the final fisheries plan to the governmental resources agencies.

TABLE 5.1. Characteristics of Weining County (Guizhou Province, China) and the Caohai Nature Reserve (CNR)

Weining County:	area	6925 km^2
	population	> 1,000,000 (in 25 townships and 600 villages)
	per capita income	2400 CNY annually
CNR area		96 km^2
Caohai Lake:	maximum flooded area	60 km^2
	present flooded area	20 km^2
	present mean depth	2 m
	present maximum depth	5 m
	present surface elevation	2171 m above MSL
	macrophyte production	1309 g/m^2

FIGURE 5.3. Crucian carp for sale in the Weining County market.

Questions

Part A: ***Assessment***

1. What level of fisheries productivity would you expect from a lake like Caohai, given its elevation, trophic state and the species that dominate the fishery?

2. What are the major water quality considerations on Caohai? What interactions likely exist between water quality and the fishery, and what considerations should be given to improving water quality?
3. Identify critical information that you need to accomplish the mission, and develop a preliminary plan for its collection.

Part B: ***Management Recommendations***

1. What recommendations would you make to improve the productivity of the Caohai fishery? Justify your answer relative to possible alternative actions.
2. What routine monitoring programs would you recommend be instituted on the lake? How would they be funded?

Part C: ***Selling the Plan***

1. How would you present your assessment results and management suggestions to the Weining County Planning Board?
2. How would you approach enlisting the support of the local fishermen for the new management plan, remembering that they have shown hostility toward fishery management efforts in the past?

References

Guizhou Academy of Sciences, Institute of Biology. 1986. Scientific survey reports on the Lake Caohai, Guizhou, China. Guizhou Nationalities Publishing House, Guiyang, Guizhou, Peoples Republic of China.

Hardin, G. 1968. The tragedy of the commons. Science 162:1243–1248.

Shouli, H., J. Harris, W. Wanying, and Y. Yongqing. 2000. Community-based conservation and development: strategies and practice at Caohai. Guizhou Nationalities Publishing House, Guiyang, Guizhou, Peoples Republic of China.

Important Conversion Factors

7 CNY (China Yuan Renminbi) = $1 USD
1 ha = 2.471 ac
1 ha = 10,000 m^2
1 km^2 = 100 ha
1 km = 0.6214 mi
1 mt (metric ton) = 1000 kg = 2205 lbs
1 kg = 2.205 lbs

NOTES

NOTES

Case 6

What Factors are Related to Condition of Flannelmouth Suckers in the Colorado River?

The Problem

The flannelmouth sucker (Figure 6.1) is endemic to the Colorado River, and remains relatively common in the Grand Canyon compared to other native fishes such as the Colorado pikeminnow that has been extirpated from this portion of the river. Your challenge will be to assess body condition (relative plumpness) data for flannelmouth sucker to familiarize yourself with this case, and then develop further questions that delve into the dynamics of this fish species and other members of the fish assemblage in the Grand Canyon.

Background

The study area included in this case study encompasses the Colorado River beginning 26 river kilometers (RKM) below Glen Canyon Dam near the Arizona-Utah border and downstream to Diamond Creek at RKM 363.2 (Figure 6.2). The study site is strongly influenced by Glen Canyon Dam, which has altered flows, reduced and stabilized water temperatures (i.e., reduced summer and increased winter temperatures), and reduced sediment transport. Flood frequency and seasonal variation in flow were greatly reduced after construction of Glen Canyon Dam. In addition, hypolimnetic releases from Glen Canyon Dam have stabilized water temperature at about 10°C (with little warming throughout the 363-km study area) as compared to pre-dam conditions when water temperature varied from near 0 to 29.4°C.

There are several primary tributaries of the Colorado River in Grand Canyon that are important to native fishes, including the Little Colorado River at RKM 98.7 and the Paria River at RKM 1.4 (Figure 6.2). These tributaries are known spawning locations for native fishes, including flannelmouth suckers, and have remained relatively unchanged with regard to flow and temperature regimes, as compared to the mainstem Colorado River.

Figure 6.1. Flannelmouth sucker *Catostomus latipinnis* collected from the Colorado River.

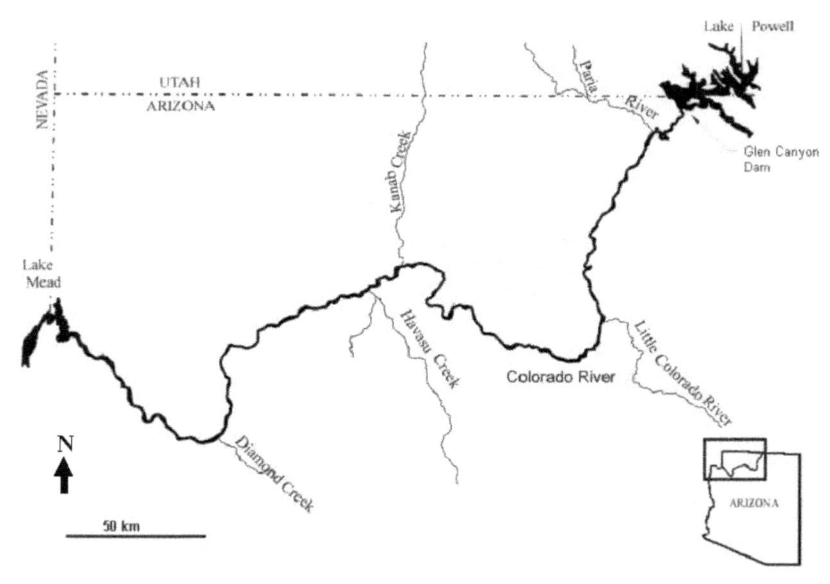

Figure 6.2. Map of the Colorado River below Grand Canyon Dam.

The condition index: Relative weight (*Wr*) is one of several indices that can be used to determine fish condition (i.e., relative plumpness). Please see Anderson and Neumann (1996) for an overview of this index and information on calculation, if necessary. One of the limiting factors for any *Wr* analysis is the availability of a standard weight (*Ws*) equation. However, Didenko et al. (2004) provided *Ws* equations for four desert fishes, including the flannelmouth sucker.

Questions

Your instructor will divide the class into a number of small teams. Each team will meet, and discuss the background information on flannelmouth suckers that is provided in Figures 6.3–6.5. Each team should prepare two questions that arise from your review of this information. A class discussion will then be based on the questions that each team has asked.

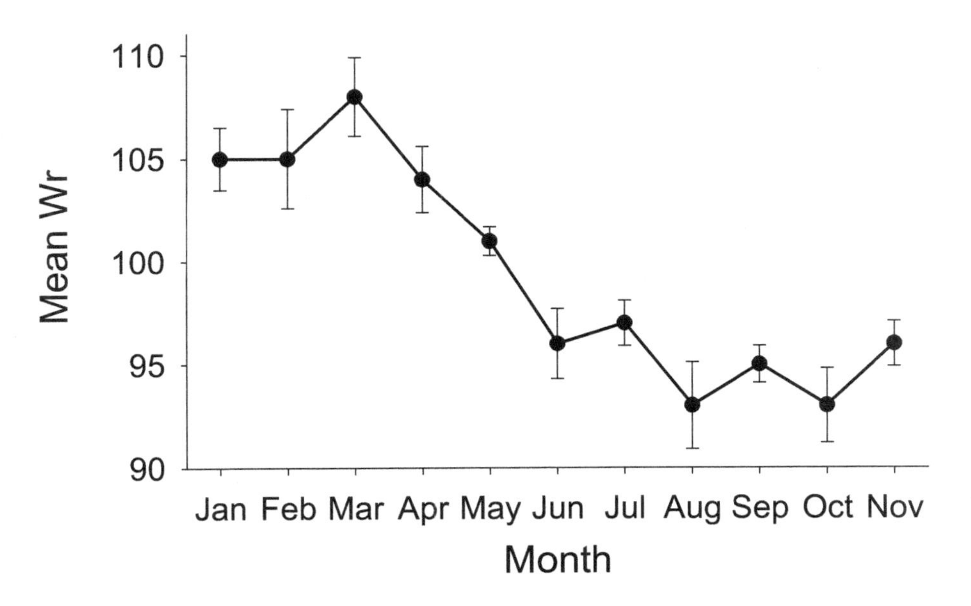

FIGURE 6.3. Mean relative weight (*Wr*)(±1 standard error) values across months for flannelmouth suckers collected in Grand Canyon, Arizona, 1993.

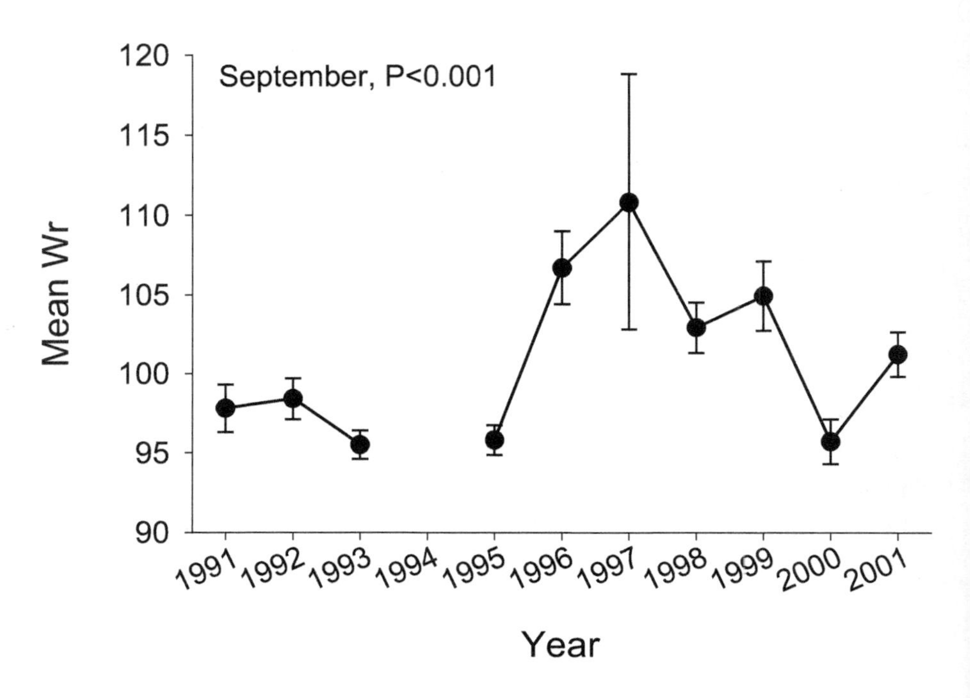

FIGURE 6.4. Mean relative weight (*Wr*)(±1 SE) values across years for flannelmouth suckers collected in Grand Canyon, Arizona in September, 1991–2001.

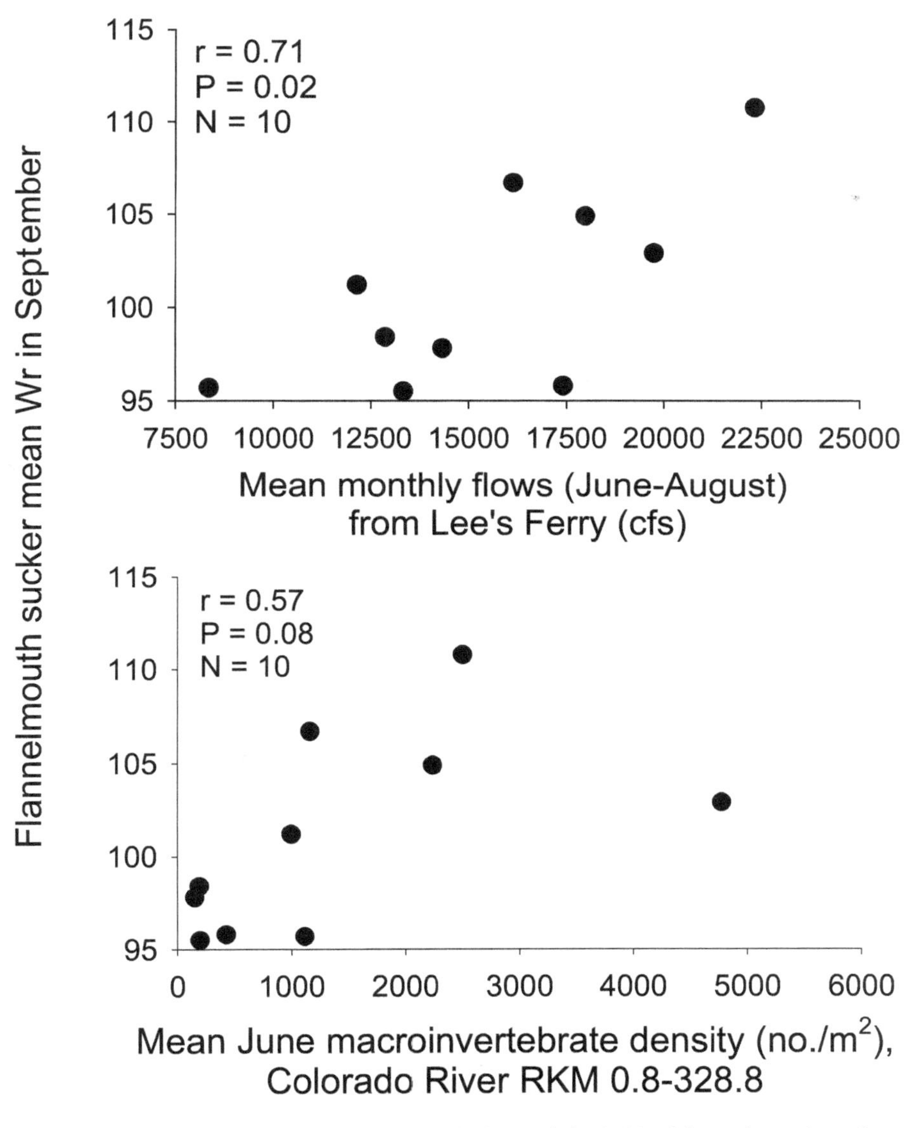

FIGURE 6.5. Relationship between mean relative weight (*Wr*) of flannelmouth suckers collected in September with mean monthly summer (June–August) flows from Lee's Ferry (at the Paria River confluence; Figure 6.1), 1991–2001 (top) and mean *Wr* of flannelmouth suckers collected in September with mean June macroinvertebrate density collected throughout the Grand Canyon (Benenati et al. 2002), river kilometers (RKM) 0.8–328.8, 1991–2001 (bottom). No fish were collected in September 1994; cfs = cubic feet per second, no. = number.

References

Anderson, R. O., and R. M. Neumann. 1996. Length, weight, and associated structural indices. Pages 447–481 *in* B. R. Murphy and D. W. Willis, editors. Fisheries techniques, 2nd edition. American Fisheries Society, Bethesda, Maryland.

Benenati, E. P., J. P. Shannon, G. A. Haden, K. Straka, and D. W. Blinn. 2002. Monitoring and research: the aquatic food base in the Colorado River, Arizona during 1991–2001. Final Report, Grand Canyon Monitoring and Research Center, U.S. Geological Survey, Cooperative Agreement Number 1452-98-FC-225590, Flagstaff, Arizona.

Didenko, A., S. A. Bonar, and W. J. Matter. 2004. Standard weight (*Ws*) equations for four rare desert fishes. North American Journal of Fisheries Management 24:697–703.

NOTES

NOTES

Case 7

Evaluating the Population Status of Black Sea Bass: One Step at a Time

Part 1

Introduction

In fisheries science, we never have perfect knowledge of a population, ecosystem, or fishery. However, we may be asked to make decisions about how to manage a stock even when we are missing some important information.

Stock status

In many marine fisheries, management is based on both the fish population abundance or biomass and the current level of fishing mortality. There are two terms that express a stock's status with respect to these two areas: overfished and overfishing. If a stock is "overfished," its biomass has been depleted below a sustainable level (i.e., the biomass that would produce Maximum Sustainable Yield). If a stock is "undergoing overfishing," the fishing mortality is higher than what is sustainable. Every stock will be within one of the four areas in Figure 7.1. For example, a previously unharvested population ($B/B_{MSY} > 1$) which just became commercially valuable and has high fishing levels ($F/F_{MSY} > 1$) would be undergoing overfishing but *not* overfished.

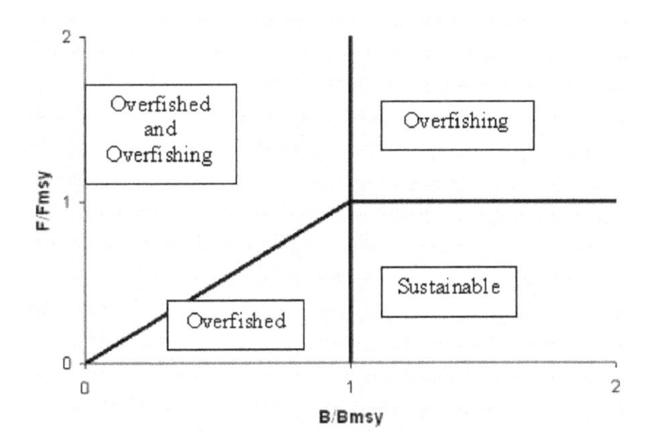

Figure 7.1. Stock status categories for marine fisheries management, based on the current biomass relative to the biomass that would produce MSY (B/B_{MSY}) and the current fishing mortality relative to the fishing mortality at MSY (F/F_{MSY}).

Black Sea Bass Facts

In this case study, you will try to piece together the population status of the black sea bass *Centropristis striata* stock off the coast of the southeastern United States. This species is important to commercial and recreational fisheries in the region; it is the 9th most commercially valuable finfish and the 4th most frequently caught finfish in the recreational fishery (NMFS, personal communication). Black sea bass are associated with sponge and coral reef areas, and individuals change sex from female to male during their lives (they are protogynous hermaphrodites).

Landings

Here is your first piece of information about black sea bass:

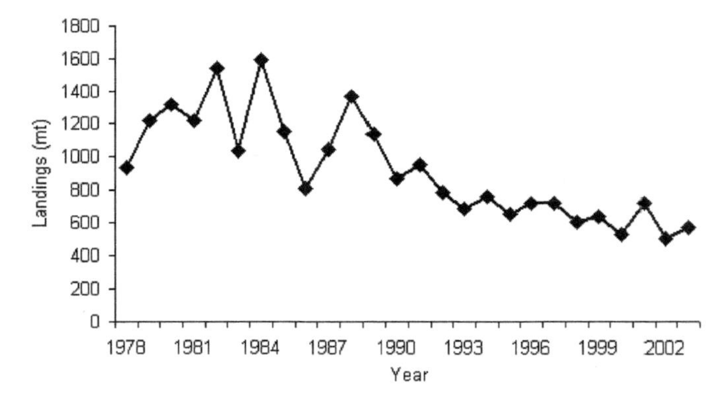

FIGURE 7.2. Black sea bass landings (in metric tons) off the southeastern U.S. coast, from commercial and recreational fisheries combined (SEDAR 2006).

Questions:

1. Summarize what this piece of information tells you.
2. What are some possible causes of a trend in landings?
3. What other information would help you identify which of these causes may be right or wrong?
4. What do you think the stock status is? How sure are you of your answer? (Shade in your best guess in the figure below.)

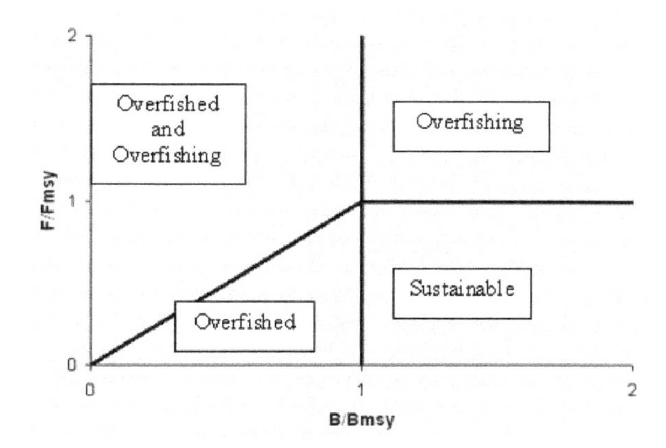

Evaluating the Population Status of Black Sea Bass: One Step at a Time

Part 2

Harvest Regulations

Here is your next piece of information about black sea bass:

TABLE 7.1. Commercial and recreational harvest restrictions for black sea bass off the southeastern U.S. coast (SEDAR 2006).

Date	Amendment	Details
31 Aug 1983	FMP	8" total length (TL) minimum size limit; 4" trawl mesh size
12 Jan 1989	1	Prohibit trawls
1 Jan 1992	4	Prohibit fish traps, entanglement nets, and longline gear within 50 fathoms; black sea bass pot gear and identification requirements
Dec 1998	8	Limited entry program; transferable permits and 225-pound non-transferable permits
24 Feb 1999	9	10" TL minimum size limit and 20 fish recreational bag limit; escape panel in traps

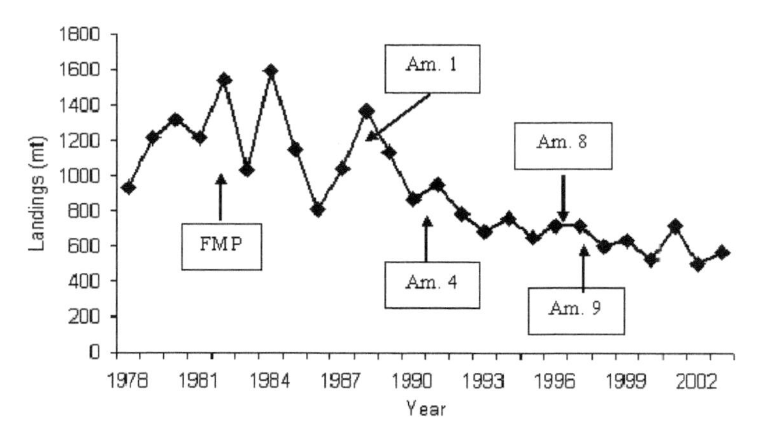

FIGURE 7.3. Black sea bass landings (same as Figure 7.2) with management actions from Table 7.1 identified.

Questions:

1. Summarize what this new piece of information tells you.
2. Do you think the regulations are responsible for the trend in landings?
3. What other information would help you determine the population status?
4. What do you think the stock status is? How sure are you of your answer?

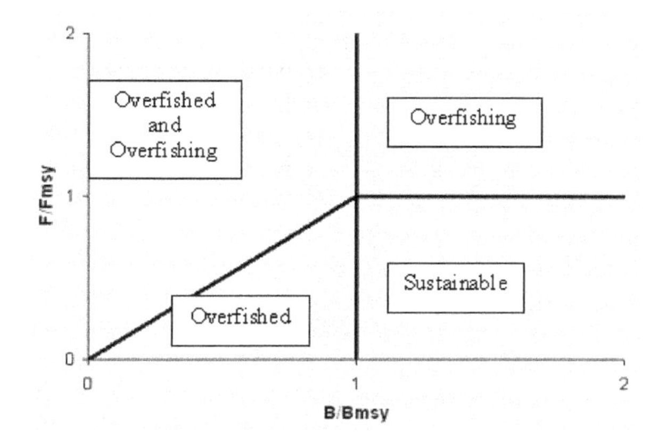

Evaluating the Population Status of Black Sea Bass: One Step at a Time
Part 3

Abundance indices (or catch-per-unit-effort)
Here is your third piece of information about black sea bass:

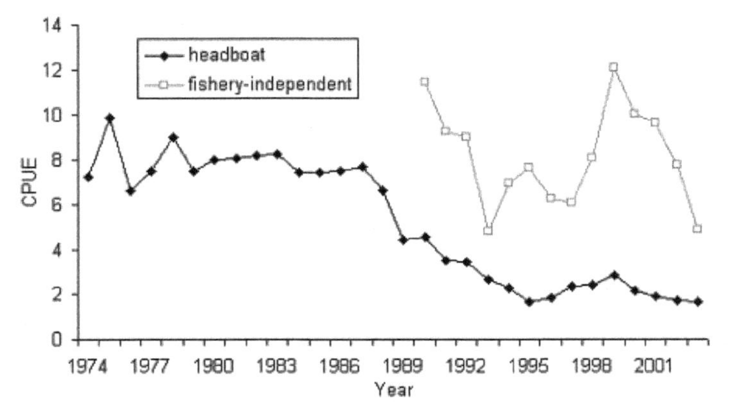

FIGURE 7.4. Abundance indices from the headboat fishery (i.e., large charter boats) and a fishery-independent survey (Chevron trap gear), showing catch-per-unit-effort over time (SEDAR 2006). Headboats are part of the recreational fishery, and headboat landings are governed by recreational fishing regulations (Table 7.1). The fishery-independent survey uses constant gear and effort and is unaffected by regulations; it is therefore believed to be more representative of the underlying population.

Questions:

1. Summarize what this new piece of information tells you.
2. What are some possible problems with using the headboat catch-per-unit-effort (CPUE) to estimate population trends?
3. What are some possible problems with using the fishery-independent CPUE to estimate population trends?
4. What other information would help you determine the population status?
5. What do you think the stock status is? How sure are you of your answer?

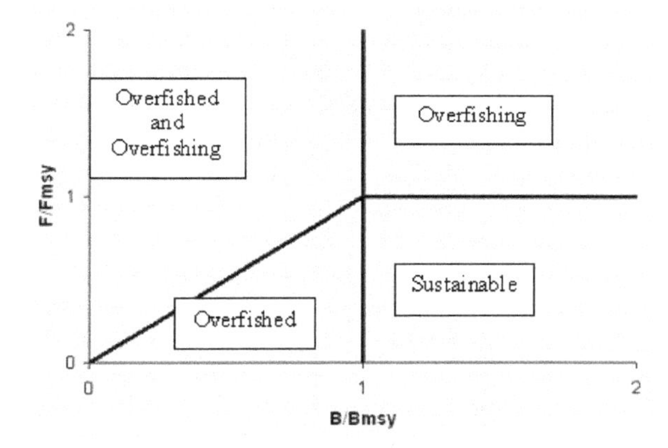

Evaluating the Population Status of Black Sea Bass: One Step at a Time

Part 4

Relative Fishing Mortality

The landings data, survey data, and other information were used in a stock assessment model to estimate fishing mortality over time. Figure 7.5 shows relative fishing mortality, which equals each year's fishing mortality divided by the fishing mortality that would produce Maximum Sustainable Yield (MSY). F/F_{MSY} is the metric used to determine whether a stock is undergoing overfishing. If $F/F_{MSY} > 1$, fishing mortality is higher than what is sustainable and the population is undergoing overfishing.

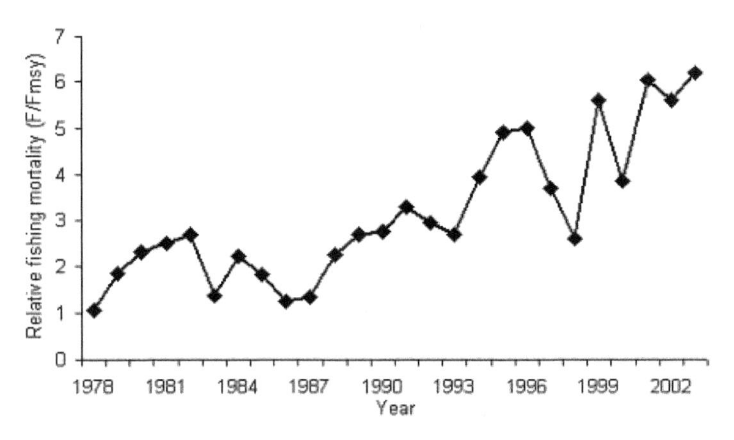

Figure 7.5. Relative fishing mortality (F/F_{MSY}) of black sea bass over time, as estimated by a stock assessment model (SEDAR 2006).

Questions:

1. Summarize what this new piece of information tells you.
2. How could relative fishing mortality increase while landings decreased?
3. Knowing that these results are from a stock assessment model, what are some possible problems with using this graph for management?
4. What other information would help you determine the population status?
5. What do you think the stock status is? How sure are you of your answer?

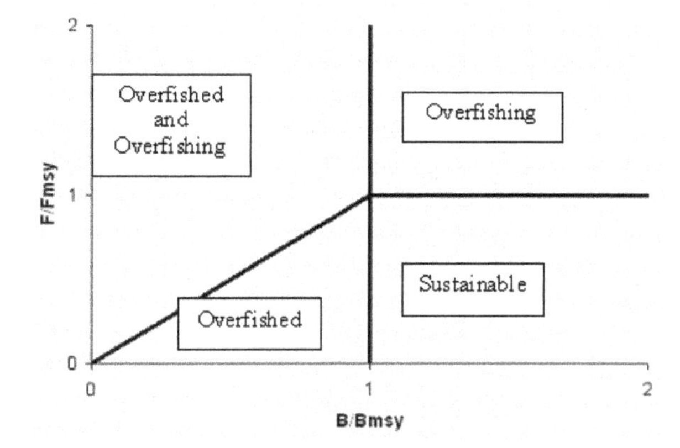

Evaluating the Population Status of Black Sea Bass: One Step at a Time

Part 5

Relative spawning stock biomass

The stock assessment model also estimated relative spawning stock biomass (SSB/SSB_{MSY}) over time. Spawning stock biomass is the estimated biomass of all mature males and females in the population. Just like for relative fishing mortality, each year's estimate is divided by the spawning stock biomass that would produce MSY to give us SSB/SSB_{MSY}. A stock is overfished if $SSB/SSB_{MSY} < 1$.

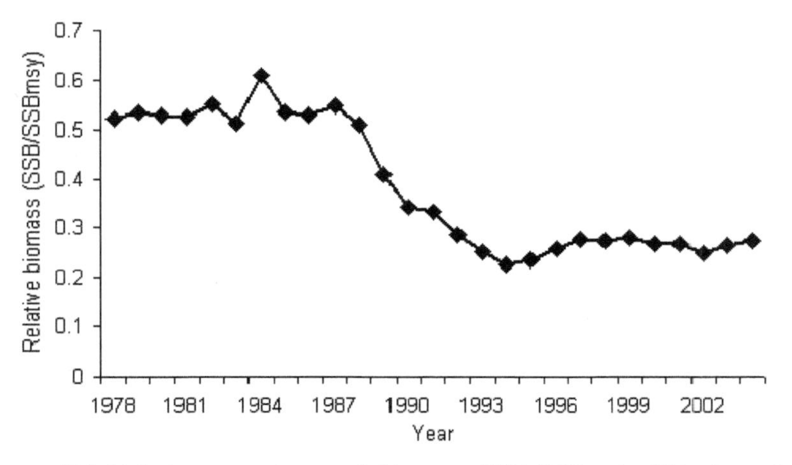

FIGURE 7.6. Relative spawning stock biomass (SSB/SSB_{MSY}) of black sea bass over time, as estimated by a stock assessment model (SEDAR 2006).

Questions:

1. Summarize what this piece of information tells you.
2. What do you think the stock status is? How sure are you of your answer?

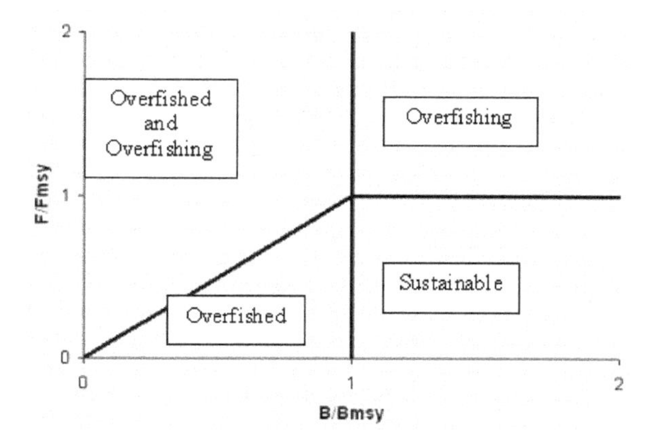

Assignment

Limiting your response to about 1 page, write your thoughts on the following questions:

1. How do you think management of this stock should proceed?
2. What factors could be affecting this black sea bass population?

FIGURE **7.7.** Black sea bass *Centropristis striata*

References

NMFS (National Marine Fisheries Service). Fisheries Statistics Division, Silver Spring, Maryland. Available: http://www.st.nmfs.gov/st1/ (October 2009).

SEDAR (Southeast Data, Assessment, and Review). 2006. Report of stock assessment: black sea bass; SEDAR update process #1. National Marine Fisheries Service Southeast Science Center, Miami, Florida. Available: http://www.sefsc.noaa.gov/sedar/ (October 2009).

NOTES

NOTES

Case 8

Predators Eat Prey: Effects of an Inadvertent Introduction of Northern Pike on an Established Fish Community

Setting

Approximately one million years ago, an area of north-central Nebraska was modified by wind to form sand hills. Approximately 1,500 natural lakes occur within lowlands in this region, and nearly half are capable of supporting fish communities. All of these lakes are shallow. Most have limited surface water drainage and are maintained by surface runoff and groundwater connections. These lakes provide some of the best angling opportunities in Nebraska. The nature of Sandhill lakes can both limit and promote fish production. Their shallow basins with clear water and abundant vegetation can result in fish kills, especially in winter. During the growing season, the basin of a Sandhill lake amounts to littoral habitat throughout. Invertebrates are abundant and fish growth rates can be fast by the standards of this geographic location. See Figure 8.1 for a view of a typical Sandhill lake.

Problem or Dilemma

Sandhill lakes are popular with both resident and nonresident anglers, with ice fishing opportunities for panfish species such as bluegill and yellow perch being especially common. Because of the high productivity and abundant zooplankton and macroinvertebrate communities in these lakes, many lakes produce exceptional sizes of bluegill (e.g., >25 cm) and yellow perch (e.g., >30 cm).

The northern pike was likely native to Sandhill lakes (Jones 1963), and this locale represents the southwestern-most native distribution for this species. As the top-level predator in many aquatic systems in both North America and Europe, northern pike can structure fish communities through predation. The purpose of this case study is to investigate the effects of a northern pike introduction on established populations of bluegill, yellow perch, and large-mouth bass.

Reading Assignment

Investigate the food habits of northern pike through library or internet research. What prey items are typically consumed by northern pike at different geographic locations? What sizes (lengths, depths) of prey will various lengths of northern pike consume? What range in biomass do northern pike populations exhibit at various locations?

Study Site

West Long Lake is a 25-ha natural lake located on the Valentine National Wildlife Refuge in Cherry County, Nebraska. Mean depth of West Long Lake is 1.2 m with a maximum depth of 1.8 m. Submergent vegetation covers approximately 80% of the lake surface area during mid-summer; however, this vegetation is interspersed throughout the lake. Grasslands completely surround the lake and there are no crops within its watershed. The long-term fish community consisted of yellow perch, bluegill, largemouth bass, and black bullhead.

Northern Pike Population Development

Northern pike were first documented in West Long Lake in Year 1 of this study, when two small pike of 34 and 37 cm (total length) were collected. In Year 5 of this study, the northern pike population (≥24 cm) was estimated at 894 fish (95% confidence interval = ±221), with density estimates of about 35.8 fish/ha (±8.8). Biomass was estimated at 22 kg/ha (±5.4). Northern pike exhibited an extended size structure by Year 5 (Figure 8.2). This northern pike population thus developed in a relatively short time period (i.e., 4 years).

The Question

In the spring of Year 1, trap nets (i.e., modified fyke nets; Hubert 1996) were used to sample yellow perch and bluegills, while night electrofishing (Reynolds 1996) was used to sample largemouth bass. The sampling data are summarized in Table 8.1.

Both panfish species had size (i.e., length) structures that were dominated by large individuals. For example, 40% of the stock-length bluegills (i.e., those 8 cm or longer) exceeded 20 cm, while 62% of the stock-length yellow perch (i.e., those 13 cm or longer) exceeded 25 cm. Such panfish populations are the source of both resident and nonresident angler interest in Sandhill lakes.

FIGURE 8.1. Nebraska Sandhill lakes tend to be shallow, windblown, and productive. Lakes such as the one pictured tend to be maintained by groundwater connection. The surrounding hills are wind-deposited sand, and the terrestrial communities are sensitive to disturbance.

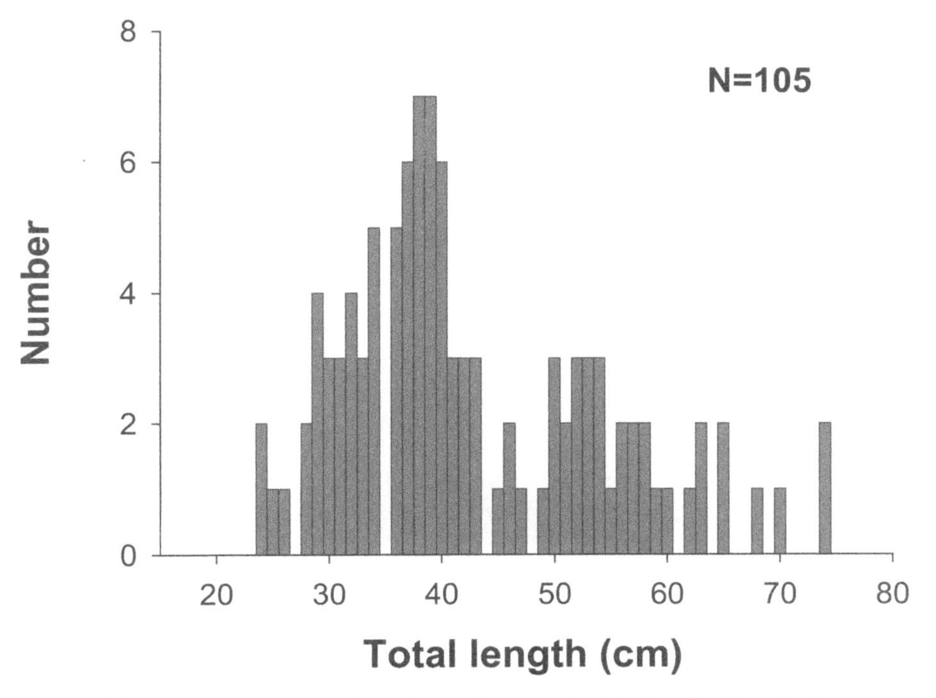

FIGURE 8.2. Length frequency of northern pike sampled during the Year 5 pike population estimate in West Long Lake, Nebraska. Sample size (N) represents the first 105 northern pike sampled with modified-fyke nets.

TABLE 8.1. Mean catch per unit effort (CPUE; number of fish ≥ stock length caught per trap-net night or per hour of night electrofishing), proportional size distribution (PSD; Guy et al. 2007), proportional size distribution of preferred-length fish (PSD-P), and mean total length (mm) at time of capture by age group for bluegill (BLG), largemouth bass (LMB), and yellow perch (YEP) sampled from West Long Lake, Nebraska. Values in parentheses for CPUE and mean length at age are standard errors; values in parentheses for PSD and PSD-P are 95% confidence intervals. Mean CPUE is based on trap-net samples for bluegill and yellow perch, and on night electrofishing samples for largemouth bass. Stock, quality, and preferred lengths for these fish species are defined in Table 8.2.

		Year 1	Year 5
BLG	CPUE ≥ stock length	28.2 (4.5)	
	PSD	68 (9)	
	PSD-P	40 (10)	
	Mean length (mm) at age 2	120.9 (3.5)	
	Mean length (mm) at age 3	156.6 (5.4)	
LMB	CPUE ≥ stock length	143.7 (32.1)	
	PSD	29 (9)	
	PSD-P	9 (6)	
	Mean length (mm) at age 3	226.6 (3.5)	
	Mean length (mm) at age 4	270.2 (9.3)	
	Mean length (mm) at age 5	304.2 (5.3)	
YEP	CPUE ≥ stock length	14.2 (2.9)	
	PSD	74 (9)	
	PSD-P	62 (10)	
	Mean length (mm) at age 2	150.5 (3.8)	
	Mean length (mm) at age 3	180.5 (5.6)	

TABLE 8.2. Length-categorization system (total length) used for fishes collected from West Long Lake, Nebraska (Gabelhouse 1984).

Species	Stock (mm)	Quality (mm)	Preferred (mm)
Bluegill	80	150	200
Largemouth bass	200	300	380
Northern pike	350	530	710
Yellow perch	130	200	250

Conversely, the size structure of the largemouth bass population was dominated by smaller individuals. For stock-length largemouth bass (i.e., those 20 cm or longer), 71% of the sample was congregated between 20 and 30 cm, and only 9% exceeded 38 cm in total length. This size structure was the result of high largemouth bass abundance and slow growth that resulted from that abundance, as indicated by the 143.7 stock-length bass caught per hour of night electrofishing.

In spring of Year 5, the fish community was again sampled with similar gears at a similar time of year. You already know that a substantial northern pike population developed in the 4 years between Year 1 and Year 5. Based on your new knowledge of predatory effects of northern pike on other fishes, predict the sampling data for the Year-5 column in Table 8.1. If possible, predict actual numerical values. If you are not comfortable with the prediction of numbers, one alternative would be to insert "lower," "similar," or "higher." Once you finish this exercise, your instructor will show you the actual sampling data for the Year-5 sample.

References

Gabelhouse, D. W., Jr. 1984. A length-categorization system to assess fish stocks. North American Journal of Fisheries Management 4:273–285.

Guy, C. S., R. M. Neumann, D. W. Willis, and R. O. Anderson. 2007. Proportional size distribution (PSD): a further refinement of population size structure index terminology. Fisheries 32:348.

Hubert, W. A. 1996. Passive capture techniques. Pages 157–192 *in* B. R. Murphy and D. W. Willis, editors. Fisheries techniques, 2nd edition. American Fisheries Society, Bethesda, Maryland.

Jones, D. J. 1963. A history of Nebraska's fisheries resources. Nebraska Game, Forestry and Parks Commission. Federal Aid in Fish Restoration Project F-4-R, Final Report, Lincoln.

Reynolds, J. B. 1996. Electrofishing. Pages 221–254 *in* B. R. Murphy and D. W. Willis, editors. Fisheries techniques, 2nd edition. American Fisheries Society, Bethesda, Maryland.

NOTES

NOTES

Case 9

Misapplication of a Minimum Length Limit for Crappie Populations: Could the Mistake Have Been Avoided?

Background

Minimum length limits for crappie populations were not implemented by the South Dakota Department of Game, Fish and Parks (SDGFP) prior to 1996. Generally, minimum length limits should be applied to low to moderate density crappie populations that have moderate to fast growth by the standards of that region and in which population size structure has been negatively affected by angler harvest. Minimum length limits should not be applied to high-density, slow-growing crappie populations (Colvin 1991; Allen and Miranda 1995). The use of such a regulation on a high-density population would compound the problem of overpopulation and slow growth by further increasing crappie density and intraspecific competition.

The size structure of black crappie and white crappie populations in Lake Alvin, South Dakota had apparently been negatively affected by excessive angler harvest. Anglers complained about the small average size of crappies harvested from the lake. Relative abundance of these populations was lower than other South Dakota crappie populations that exhibited high density and slow growth, based on catch per unit effort (CPUE) in trap (i.e., modified fyke) nets. Age and growth data indicated that the Lake Alvin crappies grew at a moderate rate near the statewide average. However, size structure of both black and white crappies collected with trap nets in Lake Alvin during the early 1990s typically were below objective ranges for a balanced population. The age structure of both species showed few fish older than age 2 in trap net samples, indicating that either the fish were harvested at a young age or natural mortality limited survival of fish in these populations. The combination of low to moderate density, moderate growth, small size, and young age structure indicated that Lake Alvin crappies may have been overharvested.

Prior to pursuing a regulation change, SDGFP biologists conducted an angler opinion survey at Lake Alvin to evaluate how receptive anglers might be to restrictive harvest regulations on crappies. Initially, anglers were asked their opinion regarding a 20-cm (8-in) minimum length limit. Of the anglers who responded, 91% were in favor. These initial respondents were resurveyed to assess their opinions regarding a 23-cm minimum (9-in) length limit; 93% of those who responded were in favor. With public support, SDGFP instituted a 23-cm minimum length limit for both species in Lake Alvin beginning 1 January 1996.

Setting

Lake Alvin is a 36.4-ha impoundment located in southeastern South Dakota and is owned and managed by SDGFP. The impoundment is eutrophic, had a moderate level of turbidity, and contained few submerged aquatic macrophytes. The prey base available to crappies primarily consisted of zooplankton and macroinvertebrates. Primary sport fishes present included black and white crappies, largemouth bass, bluegill, and channel catfish. Secondary sport fishes (i.e., those present in limited abundance) included walleye, yellow perch, northern pike, and black bullhead. Other fishes present were common carp, white sucker, and green sunfish.

Lake Alvin's close (10 km) proximity to Sioux Falls (highest human population density in South Dakota) creates the potential for high angling effort. Past creel surveys have indicated angling effort from April through September as high as 472 hours/ha.

Problem or Dilemma

For the time being, accept that the minimum length limit was not successful at improving the size structure of black and white crappies in Lake Alvin. Your instructor will provide more information to document this statement after we complete an initial exercise. Given that the regulation did not work, your assignment is to assess the pre-treatment (i.e., pre-regulation) data set, and to see if we missed any evidence that might have alerted us that the regulation was not the most appropriate choice for this situation.

Look through the sampling data collected at Lake Alvin prior to implementation of the regulation. Mean CPUE data and size structure information in the form of size structure indices are provided in Table 9.1, while condition data (relative weight) are summarized in Table 9.2. Growth data in the form of mean length at age 3 were only determined in 1995. Black crappies had a mean length at age 3 of 207 mm, compared to a statewide average of 198 mm based on 34 population samples from 25 lakes across South Dakota (Guy and

TABLE 9.1. Mean catch per unit effort (CPUE, number of fish ≥ age 1/net night) ± SE, proportional size distribution of 23-cm fish (PSD-23; percent of 13-cm and longer fish that also exceeded 23 cm; Guy et al. 2007), and mean total length (mm) ± SE at time of capture for age-3 black and white crappies collected with trap (modified fyke) nets during late June/early July 1992–1995 from Lake Alvin, South Dakota. Means are presented for the pre-regulation period (1992–1995), but age-and-growth analysis was not conducted prior to 1995.

	Black crappie			White crappie		
Year	CPUE	PSD-23	Mean length at age 3 (mm)	CPUE	PSD-23	Mean length at age 3 (mm)
1992	63 ± 16	0		63 ± 23	3	
1993	11 ± 3	0		37 ± 12	1	
1994	5 ± 1	3		10 ± 3	15	
1995	7 ± 1	3	207 ± 9.0	6 ± 2	13	220 ± 9.3
Mean	**22 ± 5**	**2**	**207 ± 9.0**	**29 ± 10**	**8**	**220 ± 9.3**

TABLE 9.2. Mean relative weight (*Wr*) by length category (SE in parentheses) for black and white crappies collected with trap nets during late June/early July 1992 to 1995 from Lake Alvin, South Dakota. SS = substock (<13 cm); S = stock (13 cm), Q = quality (20 cm), P = preferred (25 cm), M= memorable (30 cm).

	Black crappie				White crappie			
Year	SS	S-Q	Q-P	P-M	SS	S-Q	Q-P	P-M
1992	138 (8)	110 (6)	86	--	101 (11)	100 (3)	78 (5)	77 (7)
1993	--	102 (1)	88 (2)	--	--	88 (1)	81 (2)	--
1994	103 (6)	96 (4)	88 (2)	--	104 (5)	89 (3)	81 (2)	81 (2)
1995	134 (7)	106 (2)	81 (3)	--	100 (7)	97 (3)	71 (2)	77

FIGURE 9.1. Lake Alvin, South Dakota.

Willis 1995). White crappies in Lake Alvin exhibited a mean length at age 3 of 220 mm in 1995, compared with a "statewide" average of 241 mm that was based only on nine population samples from four lakes.

Does any information provided here provide you with evidence that might have prevented the improper selection of this regulation?

References

Allen, M. S., and L. E. Miranda. 1995. An evaluation of the value of harvest restrictions in managing crappie fisheries. North American Journal of Fisheries Management 15:766–772.

Colvin, M. A. 1991. Evaluation of minimum size limits and reduced daily limits on the crappie populations and fisheries in five large Missouri reservoirs. North American Journal of Fisheries Management 11:585–597

Guy, C. S., and D. W. Willis. 1995. Growth of crappies in South Dakota waters. Journal of Freshwater Ecology 10:151–161.

Guy, C. S., R. M. Neumann, D. W. Willis, and R. O. Anderson. 2007. Proportional size distribution (PSD): a further refinement of population size structure index terminology. Fisheries 32:348.

NOTES

NOTES

Case 10

CSI Fisheries: The Case of the Dead Trout

You're working the fisheries beat of your city's Crime Scene Investigation lab. After months of doing nothing but sitting at your computer playing solitaire, the phone finally rings...

"Hi. This is Bob, from Insur-o-matic. I need a fish expert."

You contemplate dashing to a phone booth to change into your cape and tights, but instead you just reply, "What can I help you with today, Bob?"

"My insurance client has a small fish pond. This summer, he had to refill a pond that he had drained, so he diverted most of the water from a small creek for about an hour. The owner of a downstream rainbow trout farm claims that this killed 25,000 of his trout, and he wants $75,000 in damages. This wasn't the first or last time that my client diverted water from the creek this summer, and there weren't any problems any other time. Was my client responsible for the fish kill? Do we need to pay the trout farm owner $75,000?"

Like any good CSI agent, you start trying to fill in the missing pieces of this case. What do you still need to know to figure out if Bob's client was solely or partially responsible for the fish kill? Is $75,000 for damages a reasonable request? Your teacher has answers to some of the questions specific to this case that were provided by the insurance agent. Other questions you will need to research yourself.

Like most investigations, we will never know all the details of what happened and why. Try not to get caught up with the *small* details. Instead, focus on key issues relating to fish kills.

When you conclude your investigation, write a brief deposition (~1 page) that the insurance agent can use if this case goes to court. This statement should include your unbiased, scientific position, as an expert, on the level of fault Bob's client had in the fish kill and why. Also discuss your standing as to whether or not $75,000 is an appropriate claim and the reasoning behind your position.

Here is the first question to get you started…

Q: When and where did the fish kill happen?

A: Sunday, August 26, 2007, near Roanoke, Virginia.

NOTES

NOTES

Case 11

To Stock or Not to Stock: That is the Question—for High Elevation Wilderness Lakes

Background

Fisheries biologists today are facing changes in societal views on use of wilderness areas, sport fisheries, and consumptive fisheries (Pister 2001). This case study will provide you with a realistic (i.e., actual) example of resource managers who were faced with tough decisions in the face of changing societal views. Wyoming biologists who manage wilderness lakes in that state are challenged by the variety of opinions that various users (and non-users!) express on appropriate management strategies.

High-elevation lakes in the Rocky Mountains often had no native fish communities, but over the years, many species of trout (Salmonidae) have been stocked to provide sport fisheries. Eight Wyoming wilderness areas (see one such area in Figure 11.1) were created by the Wilderness Act of 1964— "a place where the earth and its community of life are untrammeled by man." Generally speaking, these are considered roadless areas, with no mechanical vehicles allowed. In Wyoming, these lands have 1,202 lakes suitable for trout. As of 2003, 724 lakes contained trout populations, with some supported by natural reproduction and others supported by ongoing stocking programs; 478 lakes were fishless (could support trout, but had never been stocked).

Cutthroat trout (Figure 11.2) are native to most of the drainages in Wyoming (Figure 11.3). Other trouts, such as brook trout, brown trout, and rainbow trout also currently are present in Wyoming, but are not native (i.e., they have been introduced).

When trout are introduced to previously fishless lakes, substantial changes often occur in the aquatic community. Tadpoles and frogs can be reduced or extirpated (Bradford 1989; Bradford et al. 1993). Changes in numbers and sizes of aquatic insects and zooplankton can be substantial (Dunham et al. 2004). Changes are most likely when trout occur at high densities in small,

FIGURE 11.1. Titcomb Basin in the Bridger Wilderness, Wyoming.

FIGURE 11.2. Colorado River cutthroat trout.

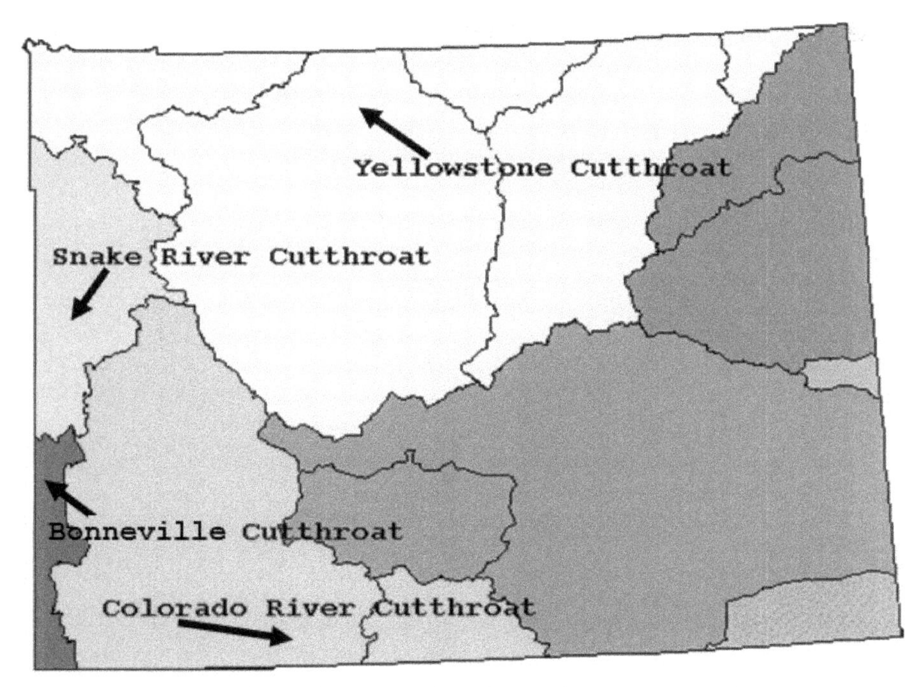

FIGURE 11.3. Four drainage basins in Wyoming with their native subspecies of cutthroat trout.

shallow lakes. Effects of trout on native organisms are more difficult to detect in larger, deeper lakes with low densities of fish.

However, anglers still request fishing opportunities. Thus, a resource manager is likely to be caught between constituent groups. Consumptive angler groups are likely to support the concept of stocking programs regardless of ecological effects. Fly fishing groups may have a more moderate position, but are likely dedicated to conservation and wise use of native trouts. In contrast, an "amphibian conservation society" will probably be concerned with the current declines in native amphibian populations across the continent, and how such populations might be affected by trout stocking. Similarly, the "defenders of invertebrates" may recognize that invertebrate communities change substantially after fish are stocked into previously fishless waters. Finally, even if stocking programs cease, the numbers of hikers, backpackers, horse riders, photographers, etc., will still continue to increase as human populations grow. What is the "nonconsumptive" effect of all these user groups?

Your Task

Assume that you are a natural resource manager working for the Wyoming Game and Fish Department. Be sure to read the mission statement of your agency below and then answer the questions that follow.

<u>Mission statement:</u> *The Wyoming Game and Fish Department is charged with providing an adequate and flexible system of control, propagation, management and protection and regulation of all wildlife in Wyoming (W.S. 23-1-301-303, W.S. 23-1-401).*

1. Will trout stockings continue or cease? Why or why not? If stockings continue, what species of trout will be used? Why?
2. Will you introduce trout into any of the lakes that currently are fishless? Why or why not?
3. Submit a written overview of your plan for resolution of this issue.

References

Bradford, D. F. 1989. Allopatric distribution of native frogs and introduced fishes in high Sierra Nevada lakes of California: implications of the negative impact of fish introductions. Copeia 1989:775–1183.

Bradford, D. F., F. Tabatabai, and D. M. Graber. 1993. Isolation of remaining populations of the native frog, *Rana mucosa*, by introduced fishes in Sequoia and Kings Canyon national parks, California. Conservation Biology 7:882–888.

Dunham, J. B., D. S. Pilliod, and M. K. Young. 2004. Assessing the consequences of nonnative trout in headwater ecosystems in western North America. Fisheries 29(6):18–26.

Eby, L. A., W. J. Roach, L. B. Crowder, and J. A. Stanford. 2006. Effects of stocking-up freshwater food webs. Trends in Ecology and Evolution 21:576–584.

Pister, E. P. 2001. Wilderness fish planting, history and perspective. Ecosystems 4:279–286.

NOTES

NOTES

Case 12

A Protected Slot Length Limit for Largemouth Bass in a Small Impoundment: Will the Improved Size Structure Persist?

The Problem

Protected slot length limit regulations (e.g., fish between 30 and 38 cm must be released; fish larger and smaller than the slot can be harvested, depending on daily creel limits) have been used to restructure high-density, slow-growing largemouth bass populations in small impoundments (Flickinger et al. 1999; Noble and Jones 1999). Here, we will detail the simulation of a 30–38 cm (12–15 in) protected slot length limit. Largemouth bass smaller than the protected range were removed annually from a privately owned pond by spring nighttime electrofishing for four years. The largemouth bass size structure increased significantly by the fifth year. The study was then terminated. No further angling occurred, and no further removal occurred. Seven years later, you are hired as a fisheries consultant by a private conservation group that has just purchased the property. Your challenge in this case study is to predict the largemouth bass population size structure in this pond 7 years after management efforts ceased.

Background

When anglers have been willing to harvest the small largemouth bass, a protected slot length limit can result in improved bass size structure (e.g., Eder 1984; Gabelhouse 1987; Novinger 1990; Neumann et al. 1994). Typically, the management goal is to reduce the density of sub-slot fish, improve growth rates, and increase size structure, often to that of a "balanced" population (e.g., proportional size distribution [PSD; Guy et al. 2007] of 40–70 [Anderson and Neumann 1996] and relative weight [*Wr*] values of 95–105 [Blackwell et al. 2000]).

The Setting

Knox Pond (Figure 12.1) is a 2.9-ha private impoundment located in the rolling prairies of western South Dakota (Jones County). During mid-summer,

FIGURE 12.1. Aerial view of Knox Pond in the rolling prairies of western South Dakota.

submergent aquatic vegetation typically covers approximately 60% of the pond surface area. The fish community includes only largemouth bass and black bullheads. The impoundment is located on a private ranch far from any town, and the largemouth bass population was unexploited prior to the study. If you are interested, find the town of Murdo, South Dakota on a web-based program such as *Google Earth.* Move 30 km north of that town, and note the remote nature of that geographic location.

Initially, the pond contained a high-density, slow-growing largemouth bass population. The biomass estimate in Year 1 was 112 kg/ha (95% confidence interval [CI] = ±21), catch per unit effort (CPUE) was 306 stock-length (i.e., ≥20 cm) largemouth bass per hour of spring nighttime electrofishing, PSD was 0, and mean *Wr* for 20–29 cm bass was 77 (95% CI = ±2). In Year 1, a 200-mm largemouth typically grew 45 mm during the next growing season.

Black bullhead abundance was low, which is not surprising given the abundance of predators and bullhead vulnerability to predation. Most black bullheads were longer than 300 mm in length because of substantial predation on the small bullheads (Saffel et al. 1990).

Details of the Removal and Subsequent Improvements in Population Size Structure

A 30–38 cm (12–15 in) protected slot regulation was simulated by removal (via electrofishing) of 20- to 30-cm largemouth bass from Year 1 through Year 4. The original goal was to annually remove 40% of the population estimate for 20-cm and longer largemouth bass, by removing only the smaller fish (i.e., <30 cm). In actuality, we removed 48, 40, 40, and 21% of the population estimates for Years 1–4, respectively. By Year 4, only 21% of the fish in electrofishing samples were <30 cm, and all fish that exceeded 30 cm were released back into the impoundment to simulate the regulation effects. Consider the following table (Table 12.1) of annual sampling data, and follow trends in size structure over time.

TABLE 12.1. Spring nighttime electrofishing data for Knox Pond, South Dakota. Density and biomass were based on Petersen mark-and-recapture population estimates. Confidence intervals and/or standard errors of the mean can be found in Neumann et al. (1994). CPUE = number of stock-length fish/hr of night electrofishing; PSD = proportional size distribution (percentage of 20-cm and longer largemouth bass that also were 30 cm or longer); PSD-P = proportional size distribution of preferred-length fish (percentage of 20-cm and longer bass that also were 38 cm or longer); S = stock length (20 cm for largemouth bass); Q = quality (30 cm); P = preferred (38 cm)(Gabelhouse 1984).

	Year 1	Year 2	Year 3	Year 4	Year 5	Year 12	Year 13
Density (number)	1,153	458	167	191	a	b	b
Biomass (kg/ha)	112	52	16	31	a		
CPUE	306	130	8	28	26		
PSD	0	0	3	47	22		
PSD-P	0	0	0	0	9		
Mean *Wr* (S-Q)	77	87	82	109	100		
Mean *Wr* (Q-P)	c	c	73	106	91		

a = population estimates not completed in Year 5 because no removal was planned
b = no population estimates were conducted
c = no fish of this size were present in the sample

A common technique for assessing populations in small impoundments is the use of a tic-tac-toe graph (also termed a *phase plane graph*; Anderson and Neumann 1996). Figure 12.2 depicts the mean *Wr* values for stock- to quality-length (20–29 cm) largemouth bass as a function of bass PSD, with all data based on the standard electrofishing surveys. The dotted lines delineate a balanced PSD of 40–70, and a *Wr* objective range of 95–105. This type of figure makes it easy to follow population changes over time.

Largemouth bass growth increased from the Year 1 to the Year 5 sample. In Year 1, a 200-mm largemouth bass averaged an annual growth increment of about 45 mm. By Year 5, average annual growth for a 200-mm largemouth bass was slightly more than 100 mm. That is a substantial increase from a bioenergetic point of view.

No further management efforts were undertaken at Knox Pond after the spring Year 5 sample, and no angler exploitation of largemouth bass occurred at the impoundment. Then, the pond was sold to the Turner Foundation as part of

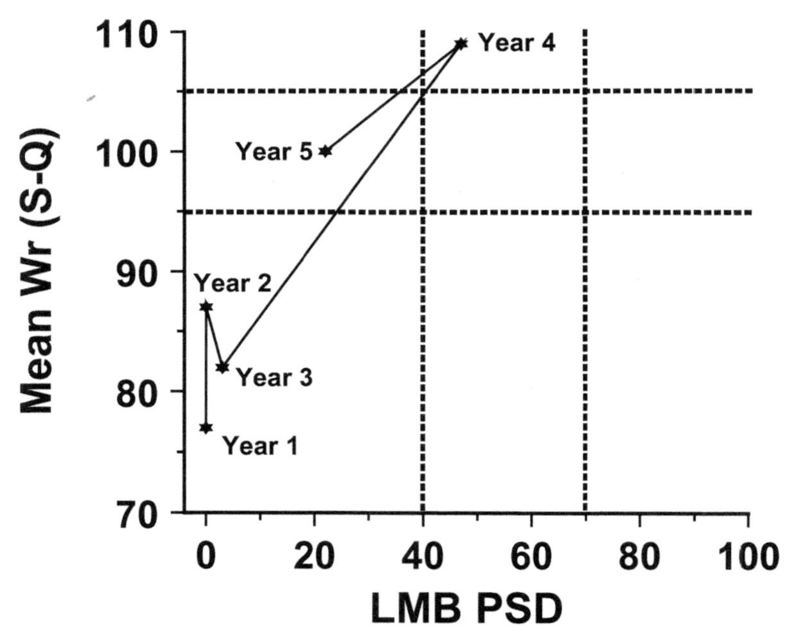

Figure 12.2. Changes in largemouth bass proportional size distribution (PSD) and mean relative weight of stock- to quality-length bass [Mean *Wr* (S-Q)] over a 5-year period in Knox Pond, South Dakota. All data were collected during spring nighttime electrofishing.

a large parcel of land. In cooperation with one of the Turner Foundation biologists, we sampled the largemouth bass during Years 12 and 13 to assess the population. Thus, we had a unique opportunity to look at this largemouth bass population 7 years after initial management efforts, and in the absence of angler effects on the population.

Questions

You have just been hired by the Turner Foundation as a fisheries specialist. Your supervisor, a wildlife biologist by education and experience, asks what you believe happened to the Knox Pond largemouth bass population since the time of the initial study (remember, 8 years have passed since the last removal and sampling efforts). Consider the probable spring nighttime electrofishing data from Years 12 and 13. What do you expect for size structure (e.g., PSD), catch per unit effort, and condition (*Wr*) indices? Why? On your copy of Table 1, predict the likely values for each sampling index in the empty cells. Be sure to explain why you expect that particular change. NOTE: no density and biomass estimates were conducted in years 12 and 13.

References

Anderson, R. O., and R. M. Neumann. 1996. Length, weight, and associated structural indices. Pages 447–481 *in* B. R. Murphy and D. W. Willis, editors. Fisheries techniques, 2nd edition. American Fisheries Society, Bethesda, Maryland.

Blackwell, B. G., M. L. Brown, and D. W. Willis. 2000. Relative weight (*Wr*) status and current use in fisheries assessment and management. Reviews in Fisheries Science 8:1–44.

Eder, S. 1984. Effectiveness of an imposed slot length limit of 12.0–14.9 inches on largemouth bass. North American Journal of Fisheries Management 4:469–478.

Flickinger, S. A., F. J. Bulow, and D. W. Willis. 1999. Small impoundments. Pages 561–587 *in* C.C. Kohler and W. A. Hubert, editors. Inland fisheries management in North America, 2nd edition. American Fisheries Society, Bethesda, Maryland.

Gabelhouse, D. W., Jr. 1984. A length-categorization system to assess fish stocks. North American Journal of Fisheries Management 4:273–285.

Gabelhouse, D. W., Jr. 1987. Responses of largemouth bass and bluegills to removal of surplus largemouth bass from a Kansas pond. North American Journal of Fisheries Management 7:81–90.

Neumann, R. M., D. W. Willis, and D. D. Mann. 1994. Evaluation of largemouth bass slot length limits in two small South Dakota impoundments. The Prairie Naturalist 26:15–32.

Noble, R. L., and T. W. Jones. 1999. Managing fisheries with regulations. Pages 455–477 *in* C. C. Kohler and W. A. Hubert, editors. Inland fisheries management in North America, 2nd edition. American Fisheries Society, Bethesda, Maryland.

Novinger, G. D. 1990. Slot length limits for largemouth bass in small private impoundments. North American Journal of Fisheries Management 10:330–337.

Saffel, P. D., C. S. Guy, and D. W. Willis. 1990. Population structure of largemouth bass and black bullheads in South Dakota ponds. The Prairie Naturalist 22:113–118.

NOTES

NOTES

Case 13

Horseshoe Crabs: A Struggle Among User Groups

Late Triassic Period (about 200 million years ago), at what will someday be the East Coast of the United States:

> *The mountains far in the distance are jagged and steep, having only recently broken through the earth. The climate is quite warm and the air smells slightly like methane. A large dinosaur pushes slowly through the cycad trees, flattening ferns with its long tail. Like other prosauropods, this 30-foot long Ammosaurus has a long neck, allowing it to graze high in the tree tops. A small group of theropods dash past on the mudflats, hungrily eying the Ammosaurus. Luckily for the grazer, it is too large for this group of skilled hunters to eat (Weishampel and Young 1996). As night falls and the tide rises, another strange creature emerges from the water. Thousands of animals that resemble helmets crawl onto the beach to spawn before returning to the depths of the ocean. These horseshoe crabs pay no attention to the dinosaurs; this species was here about 50 million years before the dinosaurs appeared.*

For 250 million years, horseshoe crabs *Limulus polyphemus* have conducted their annual pilgrimage to spawn on sandy beaches, particularly in the Delaware Bay. But, in recent years, their numbers have been dwindling, worrying the multiple groups that rely on this species. On the Atlantic coast of the United States, horseshoe crabs are commercially harvested for bait, are used by the biomedical industry, and are an important source of food for a large number of species including migrating shorebirds (Berkson and Shuster 1999).

Management of horseshoe crabs is an attempt to balance the requirements of these user groups. However, the needs of these groups are often difficult to reconcile, and value judgments and compromises may be needed. This case study describes the needs of these primary stakeholders in more detail and includes position statements about horseshoe crab management in Virginia from a member of a conservation advocacy group and a member of the seafood packing industry.

FIGURE 13.1. Horseshoe crabs *Limulus polyphemus*

Shorebirds and Conservationists

> *"We couldn't believe our eyes ... the flats were alive with birds. You couldn't even see the beach. It was shoulder to shoulder with bobbing, rippling, feeding shorebirds. Most were chestnut-colored red knots, but occasionally you'd also see the crossed scissor wings of laughing gulls poking into the air. The plaintive whistles of the shorebirds were matched by the raucous calls of the gulls. There must have been a hundred thousand birds and a million more beyond our line of vision."* (Sargent 2002, 74)

Every spring, hundreds of thousands of migrating shorebirds stop in the Delaware Bay during their migration from wintering grounds in North America, Central America, and South America to breeding grounds in the Arctic (Botton et al. 1994). Their stop-over coincides both spatially and temporally with the annual horseshoe crab breeding, when the crabs arrive en masse to spawn on sandy beaches (Botton and Harrington 2003). A number of shorebird species stop in the Delaware Bay, but perhaps the most well known is the red knot *Calidris canutus*. This pigeon-sized bird migrates from Tierra del Fuego at the southernmost tip of South America to Arctic breeding grounds, a one-way trip of about 10,000 miles (USFWS 2003). During a red knot's 2–3 week pit-stop in the Delaware Bay, a single bird can consume about 18,000 horseshoe crab eggs per day and increase in weight by 40–50% (USFWS 2003).

The abundances of red knots and other shorebird species have decreased since the 1980s (Clark et al. 1993). The cause of this decline is unknown, but a number of factors could be playing a role, including disease, human disturbance, heavy metal contamination, or changes to the breeding or wintering ground conditions. Declining horseshoe crab abundance and egg availability have also been identified as factors that could contribute to shorebird declines. While most of these factors are out of our control (or harder to manage; i.e., development and disturbance), conservation groups have lobbied extensively to increase horseshoe crab abundance by reducing or eliminating commercial harvest of horseshoe crabs. Birdwatchers and conservationists have become a powerful force with the increasing importance of ecotourism to the economy of the Delaware Bay region. An estimated 6,000 to 10,000 recreational birders visit the Delaware Bay in spring, bringing $6.8 to $10.3 million to the regional economy (USFWS 2003).

Horseshoe crabs also provide an important food source for the threatened Loggerhead sea turtle *Caretta caretta*. Loggerheads are one of the few animals that prey on adult horseshoe crabs, and horseshoe crabs are its most common prey item in the Chesapeake Bay area (Lutcavage and Musick 1985).

Box 13.1. Testimony before the Virginia Marine Resources Commission, 27 June 2006

Perry Plumart
Director of Conservation Advocacy
American Bird Conservancy

Virginia at the May 2006 Atlantic States Marine Fisheries Commission (ASMFC) recognized the importance of the horseshoe crab to the Red Knot and other migrating shorebirds, particularly in Delaware Bay. At the meeting, the Virginia Marine Resources Commission (VMRC) offered the regulations proposed here to attempt to reduce the number of horseshoe crabs landed in Virginia that are of Delaware Bay origin.

The ASMFC has been imposing increased regulation of horseshoe crab take because of the heavy overfishing of horseshoe crabs in the 1990s where millions of horseshoe crabs were landed in a fishery that targeted breeding females. According to a peer reviewed paper by Michelle Davis of Virginia Tech [(Davis et al. 2006)], this massive overfishing removed 60% to 80% of the entire biomass of horseshoe crabs in Delaware Bay. The result was a steep decline in the abundance of horseshoe crab eggs on the spawning beaches.

The Red Knot, a migratory shorebird, is teetering on the brink of extinction according to the world's foremost shorebird scientists. Its numbers have plummeted by 90% from the early 1990s to a population today of approximately 15,000 birds. This decline of the Red Knot directly tracks the spike in the 1990s landings of millions of horseshoe crabs.

Box 13.1. Continued.

The Red Knot and other migratory shorebirds need a superabundance of horseshoe crab eggs on the beach in order for them to gain weight and continue their epic journey to the Arctic breeding grounds. Studies have shown a decline in horseshoe crab egg availability from average counts of 40,000 to 100,000 eggs per square meter in 2000 to fewer than 5,000 eggs per meter today. The stomach contents of a Red Knot are 95% horseshoe crab eggs indicating the eggs critical importance to the bird.

The conservation oriented action offered today represents an effort by Governor Kaine's Administration and Virginia to begin to balance commercial fishing interests with wildlife conservation priorities. It represents a break from Virginia's previous policy of opposing virtually all horseshoe crab conservation regulations.

American Bird Conservancy recommends that the VMRC seek funding from the Virginia General Assembly to enforce and scientifically monitor the new regulations. The new regulations will put increased pressure on the Virginia and Chesapeake Bay populations of horseshoe crabs.

Sound scientific monitoring will work to prevent overfishing of Virginia's horseshoe crabs. Furthermore, the relationship between Virginia's horseshoe crabs and migrating shorebirds is not well understood and research should be funded to remedy this situation.

American Bird Conservancy recommends that the VMRC annually review these regulations based on the science and landings data. The VMRC needs to ensure that overfishing is not the result of the increased landings of horseshoe crabs concentrated from Virginia's waters.

In the mid-Atlantic, American Bird Conservancy continues to advocate for strong conservation measures for horseshoe crabs both at the state level and at the ASMFC. This includes strong support for a moratorium on horseshoe crab take in Delaware Bay as proposed by the States of New Jersey and Delaware.

Mr. Chairman, there is a conservation crisis in Delaware Bay and the mid-Atlantic. The Red Knot is in real danger of extinction. I urge that the new regulations considered here today be carefully monitored to ensure we have healthy horseshoe crab populations for fishing interests and migratory shorebirds.

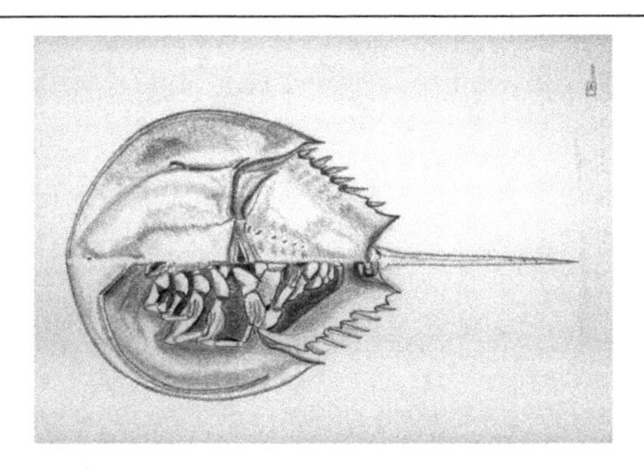

The Commercial Bait Fishery

> *"When people say that fishing is just a job, it just doesn't work. To be a fisherman, it's more a passion than the desire for the work... And when you have a passion for something, no matter what you go through, it doesn't bother you... A recent study issued by the University of Maine calculates that each fisherman generates 6.6 jobs shoreside. Every commercial boat is like a shop down on Main Street. We've got fuel bills; we've got to pay our suppliers for our gear; we've got to pay crew and maintenance. Proportionally, we spend more money than the average business, and people don't realize that. The average person thinks boats come and go and don't help support anyone else except the fisherman."* (Playfair 2003, 48–49)

From the 1850s to 1920s, horseshoe crabs were harvested in large numbers for use as fertilizer and chicken feed. In the 1870s four million crabs were harvested each year, and this number declined steadily until the fishery collapsed. The practice of collecting horseshoe crabs for fertilizer stopped in 1960, as much because of the horrible smell it caused as for concern for the horseshoe crab population (Sargent 2002). The horseshoe crab population rebounded quickly in the absence of this intensive harvest.

By the 1980s, some of the big east coast fisheries like cod and groundfish were collapsing, so the Department of Commerce encouraged fishermen to catch "underutilized" species, such as American eel *Anguilla rostrata* and conch (Family Melongenidae). Since the most effective bait for catching eels and conch in pots was horseshoe crabs, the horseshoe crab bait fishery was born and crabs were harvested by trawling, by dredging, or by hand during spawning. When the horseshoe crab bait fishery began, there were few restrictions and no harvest reporting requirements; the maximum reported coastwide harvest of about 2.7 million crabs occurred in 1998 (ASMFC 2004). A number of harvest restrictions have been put in place since then, including state-by-state quotas and spawning season closures. Additionally, the development of bait-saving devices used in eel and conch pots extended the fishing time of the bait, reducing the demand for horseshoe crabs. In 2004, the commercial bait fishery harvested fewer than 700,000 crabs coastwide, a 75% reduction from 1998 landings.

The horseshoe crab bait fishery directly and indirectly impacts a large number of people and contributes substantially to the regional economy. Conch and eel are increasingly popular seafood options, especially in overseas markets. Some of the eels are used as bait in recreational and commercial fisheries for other species, such as striped bass *Morone saxatilis*.

Box 13.2. Horseshoe Crab Economics From a Fisheries Perspective

Richard B. Robins, Jr.
Export Sales Manager
Chesapeake Bay Packing, LLC

Horseshoe crabs are uniquely valuable, both ecologically and economically. Horseshoe crab eggs constitute the primary food resource for migratory shorebirds as they make their annual stopover on their northward migration in Delaware Bay. Horseshoe crabs also support the conch (whelk, *Busycon canaliculatum*) pot fishery which ranges from Virginia to Massachusetts.

The conch fishery supports 270 to 370 jobs and generates $11 to S15 million of economic revenue (Manion et al. 2000). While the harvest of conch ranges from Virginia to Massachusetts, most of the processing industry is centered in Virginia. Three conch processing facilities are located on the Eastern Shore of Virginia, while two others are located in Newport News and Hampton. All of the plants are located in Economic Enterprise Zones. Virginia's conch processing plants generate $9 million of direct revenue, employing approximately 110 workers, and supporting approximately 75 fishing jobs in the Commonwealth. Many of the jobs are located on the Eastern Shore, parts of which are economically disadvantaged. The largest plant, for example, employs 40 to 50 workers and is located in Cheriton, Virginia, where the median household income in 2000 was $11,670 (U.S. Census 2000).

At the center of the fisheries economic issue in the current case is the fact that the conch fishery is uniquely dependent on horseshoe crabs for bait. The fishery is a pot fishery and no alternative baits have been proven to be effective—or economically viable—for the conch pot fishery. In 2000, Virginia was the first state to require the use of bait-saving devices, known as bait bags, in its conch fishery. Other mid-Atlantic states have since followed suit, and this conservation initiative has substantially reduced the industry's consumption of horseshoe crabs coastwide. Since the late 1990's, the ASMFC has cut coastwide landings of horseshoe crabs by 78%, and bait bags have enabled the conch fishery to survive economically. This regulation was the result of research by VIMS [Virginia Institute of Marine Science], and VIMS is continuing to research alternative baits and bait-saving gear in conjunction with industry.

Successful management of the horseshoe crab resource from the perspective of the fishing industry is defined as balanced management. The stakes in the management of this species are extraordinarily high, both ecologically and economically. Recognizing the ecological importance of the horseshoe crab resource for migratory shorebirds, and the apparent, immediate needs of the red knot population, the conch industry supported a limited, delayed male-only harvest of horseshoe crabs for a two year period in Delaware Bay, which was adopted as the centerpiece of Addendum IV to the ASMFC's Interstate Management Plan for Horseshoe Crabs in May, 2006. This measure prioritizes the needs of migratory shorebirds by maximizing female escapement from harvest, which, in turn, will increase egg availability for the benefit of shorebirds, while allowing for continued, limited use of the resource by the fisheries.

Box 13.2. Continued.

Addendum IV, as approved, represents an uncompromisingly risk-averse action for the benefit of migratory shorebirds, and it represents an equitable approach to a remarkably complex management issue. Multiple surveys show significant improvements in juvenile and sub-adult horseshoe crab abundance in the Delaware Bay area, and it is widely expected that horseshoe crab spawning will improve over the next several years. As an interim management strategy, Addendum IV will contribute to increased spawning activity and improved conditions for migratory shorebirds in Delaware Bay while research on alternative baits continues.

The Biomedical Industry

> *"That first year Frederik [Bang] worked with many animals. But the first time he injected some bacteria into a horseshoe crab, he knew he had discovered his research animal. You see, it did the most amazing thing. It simply sludged up and died... Well, most people would have simply thrown the crab away and picked up another. But Frederik realized something special was going on. Here you had this ancient creature that didn't have an immune system—it lacked both T cells or B cells—but it seemed to be mounting some kind of defense. Frederik knew he had stumbled onto a simple model for the human immune system."* (Sargent 2002, 34)

What Frederik Bang had discovered in the 1950s was *Limulus* amoebocyte lysate, or LAL. Cells in the horseshoe crab's blue, copper-based blood produce LAL as a defense mechanism against Gram-negative bacteria, which are ubiquitous in the benthic marine environment (Novitsky 1984). When blood is exposed to these bacteria, LAL causes the blood to clot, thereby isolating the bacteria before it can do further harm. Prior to this finding, biomedical and pharmaceutical companies tested the sterility of vaccines, syringes, scalpels, etc., on live rabbits. The discovery of LAL provided a faster, more sensitive test for contamination. The Food and Drug Administration now requires that every injectable or implantable device be tested with LAL (Sargent 2002).

Horseshoe crabs are caught by the biomedical companies, bled until a clot forms, and then released at the capture site. There is an estimated mortality of 7.5% from the bleeding process (Walls and Berkson 2003), with more of that mortality attributable to transporting the crabs than bleeding them (Hurton 2003). Approximately 280,000 horseshoe crabs were bled for LAL in 2000 (ASMFC 2004). The regional economic value of the horseshoe crab biomedical industry is $26.7 to $34.9 million annually (USFWS 2003), and the processed blood is worth over $15,000 a quart (Sargent 2002). Currently, there is no substitute for LAL that offers comparable speed and sensitivity, and horseshoe crabs are the only source of LAL.

FIGURE 13.2. Horseshoe crabs are collected for the biomedical industry in Ocean City, Maryland.

Current Horseshoe Crab Management

The horseshoe crab management board of the Atlantic States Marine Fisheries Commission (ASMFC) coordinates management and conservation of nearshore horseshoe crab fishery resources. All United States coastal Atlantic states (Maine to Florida) are members, and member states may impose stricter harvest restrictions than those required by ASMFC. The horseshoe crab management board is composed of members of state and federal agencies, state legislation, and fishing and seafood packing industries. Conservation groups and the public have opportunities to provide their input on proposed management throughout the management process, although they are not voting members of the management board.

Very few data are available to assess or manage horseshoe crabs, since reporting commercial horseshoe crab landings has only been mandatory since 1998 and only two fishery-independent surveys specifically monitor this species. The ASMFC's goal for horseshoe crab management, as stated in the 1998 Fishery Management Plan, is:

> "to conserve and protect the horseshoe crab resource to ensure its continued role in the ecology of the coastal ecosystem, while providing for continued use over time. Specifically, the goal includes

> management of horseshoe crab populations for continued use by: current and future generations of the fishing and non-fishing public (including the biomedical industry, scientific, and educational research); migrating shorebirds; and other dependent fish and wildlife, including federally listed (threatened) sea turtles" (ASMFC 1998, 20–21).

This goal represents a form of ecosystem-based management, in which two or more species are managed together. Much of fisheries management used in the past has focused on optimizing the harvest of a single species, while essentially ignoring predator-prey interactions and other ecosystem processes (Pikitch et al. 2004). Ecosystem-based management attempts to incorporate these interactions into assessment and management.

For horseshoe crabs, managers hope to increase shorebird abundance by increasing horseshoe crab abundance. However, this approach assumes a direct cause-effect link between these factors. Many other potential factors could be affecting shorebird numbers, and there are high levels of uncertainty surrounding both horseshoe crab and shorebird abundance estimates. Additionally, the management goal states the objective of also providing horseshoe crabs for harvest and the biomedical industry. Horseshoe crab managers face the challenge of balancing these goals to effectively manage horseshoe crabs for multiple, diverse user groups.

Discussion Questions

1. Is there a way to determine the "value" of providing horseshoe crabs for each user group (economic value, importance to human health, species conservation, providing jobs, providing seafood, etc.)? Should all user groups be valued equally?
2. What management approach would you recommend to balance the needs of these groups? Why do you think this is an appropriate approach?
3. Should biomedical harvest be subject to same rules as commercial bait harvest (i.e., seasonal closures, moratoria, etc.)? Why or why not?
4. Is there a conflict of interest in the management process when stakeholders serve on management boards and help determine harvest regulations? Who (and to what extent) should be involved in the management process?
5. How does ecosystem-based fishery management apply to this situation? What are some obstacles to using this approach?

References

ASMFC (Atlantic States Marine Fisheries Commission). 1998. Interstate fishery management plan for horseshoe crab. ASMFC, Fisheries Management Report 32, Washington, D.C.

ASMFC. 2004. Horseshoe crab 2004 stock assessment report. ASMFC, Washington, D.C.

Berkson, J., and C. N. Shuster, Jr. 1999. The horseshoe crab: the battle for a true multiple-use resource. Fisheries 24:6–11.

Botton, M. L., and B. A. Harrington. 2003. Synchronies in migration: shorebirds, horseshoe crabs, and Delaware Bay. Pages 5–26 *in* C. N. Shuster, R. B. Barlow, and H. J. Brockman, editors. The American Horseshoe Crab. Harvard University Press, Cambridge, Massachusetts.

Botton, M. L., R. E. Loveland, and T. R. Jacobsen. 1994. Site selection by migratory shore birds in Delaware Bay, and its relationship to beach characteristics and abundance of horseshoe crab (*Limulus polyphemus*) eggs. Auk 111:605–616.

Clark, K. E., L. J. Niles, and J. Burger. 1993. Abundance and distribution of migrant shorebirds in Delaware Bay. Condor 95:694–705.

Davis, M. L., J. Berkson, and M. Kelly. 2006. A production modeling approach to the assessment of the horseshoe crab (*Limulus polyphemus*) population in Delaware Bay. Fishery Bulletin 104:215–225.

Hurton, L. 2003. Reducing post-bleeding mortality of horseshoe crabs (*Limulus polyphemus*) used in the biomedical industry. Master's Thesis, Virginia Tech, Blacksburg, Virginia.

Lutcavage, M., and J. A. Musick. 1985. Aspects of the biology of sea turtles in Virginia. Copeia 1985:449–456.

Manion, M. M., R. A. West, R. E. Unsworth. 2000. Economic assessment of the Atlantic Coast horseshoe crab fishery, prepared for Division of Economics, U.S. Fish and Wildlife Service.

Novitsky, T. J. 1984. Discovery to commercialization: the blood of the horseshoe crab. Oceanus 27:13–18.

Pikitch, E. K., C. Santora, E. A. Babcock, A. Bakun, R. Bonfil, D. O. Conover, P. Dayton, P. Doukakis, D. Fluharty, B. Heneman, E. D. Houde, J. Link, P. A. Livingston, M. Mangel, M. K. McAllister, J. Pope, and K. J. Sainsbury. 2004. Ecosystem-based fishery management. Science 305:346–347.

Playfair, S. R. 2003. Vanishing species: saving the fish, sacrificing the fisherman. University Press of New England, Lebanon, New Hampshire.

Sargent, W. 2002. Crab wars: a tale of horseshoe crabs, bioterrorism, and human health. University Press of New England, Lebanon, New Hampshire.

U.S. Census Bureau. Census 2000, Summary File 3 (SF-3), U.S. Census Bureau, Washington, D.C.

USFWS (U.S. Fish and Wildlife Service). 2003. Delaware Bay shorebird-horseshoe crab assessment report and peer review. Migratory Bird Publication R9-03/02. USFWS, Arlington, Virginia.

Walls E. A., and J. Berkson. 2003. Effects of blood extraction on horseshoe crabs (*Limulus polyphemus*). Fishery Bulletin 101:457–459.

Weishampel, D. B., and L. Young. 1996. Dinosaurs of the East Coast. Johns Hopkins University Press, Baltimore, Maryland.

NOTES

NOTES

Case 14

Interpreting the Size Structure of a Fish Population Sample: What Can We Infer about the Dynamic Rate Functions?

Background

Three dynamic rate functions, recruitment, growth, and mortality, interact to determine fish population size and age structure. Recruitment adds new individuals to a population, which is balanced by mortality. Mortality can be due to natural causes such as predation or disease, or due to harvest by human anglers. If you are not comfortable with the definitions for recruitment, growth, and mortality, please refer to Box 1.1 in Willis and Murphy (1996). If you would like more information, but still in a general format, review Chapter 3 in Willis et al. (2009).

One of the first things that most biologists do after completing a sampling trip is to construct a length-frequency histogram for fish species of interest. Simply by assessing this histogram, some initial suppositions (i.e., educated guesses) can be made concerning the population. Is the size structure dominated by small fish? If so, is there high mortality, either from natural causes or from angling, that is causing a loss of large fish? Alternatively, could this be a high density, slow growing population that is sometimes referred to as "stunted?" What other sampling will I need to do to determine the true recruitment, growth, and mortality patterns for this population? In this exercise, we would like you to start asking such questions.

Setting

You actually need very little information to begin asking questions about the size structure of a population, and the reasons for that size structure. Here, we provide you with a length-frequency histogram for yellow perch sampled from a northern lake using an experimental (i.e., multiple-mesh) gill net (Figure 14.1). If you are unfamiliar with this sampling gear, please see Hubert (1996) for background information. This was a spring sample, so no age-0 fish will be included on this histogram; all fish are age-1 and older.

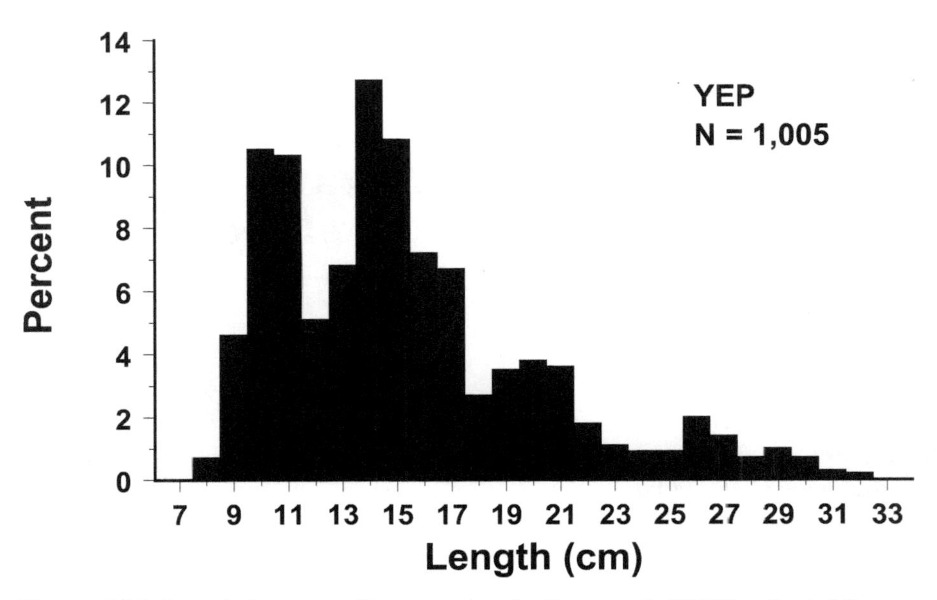

FIGURE 14.1. Length frequency for a sample of yellow perch (YEP) collected from a natural lake in the northern United States during the spring. All length measurements were maximum total length.

The Question

Your instructor will break the class into small discussion groups. Consider the length frequency of the yellow perch in this sample. Try to guess what you can about recruitment patterns. Do you believe that recruitment (i.e., year-class strength) is consistent or inconsistent? What about growth and mortality rates? Even though you do not know growth rates, based on the maximum sizes of fish in the sample, do you believe that growth is most likely 1) moderate to fast or 2) slow by the standards of that geographic location? Do you see any evidence of excessive mortality, either for total annual mortality or for angler-induced mortality?

References

Hubert, W. A. 1996. Passive capture techniques. Pages 157–192 *in* B. R. Murphy and D. W. Willis, editors. Fisheries techniques, 2nd edition. American Fisheries Society, Bethesda, Maryland.

Willis, D. W., and B. R. Murphy. 1996. Planning for sampling. Pages 1–15 *in* B. R. Murphy and D. W. Willis, editors. Fisheries techniques, 2nd edition. American Fisheries Society, Bethesda, Maryland.

Willis, D. W., C. G. Scalet, and L. D. Flake. 2009. Introduction to wildlife and fisheries: an integrated approach, 2nd edition. W. H. Freeman and Company, New York.

NOTES

NOTES

Case 15

The Debate Over Shark Abundance

There is a knock at your office door, and your colleague John pops his head in. His face is flushed, and he is clearly frustrated about something. "You got a minute?" he asks. You invite him in and prepare for his in-depth critique of last night's football game, but he is all business today and jumps right in to the purpose of his visit.

"I don't know which one to believe," John groans. "How is the management board supposed to make a decision about this?"

Your blank stare encourages him to explain the problem further. He tosses two articles on your desk and begins. "It's sharks. As a member of the International Commission for the Conservation of Atlantic Tuna (ICCAT), I help advise decision makers on how to regulate and monitor shark harvest. We try to base our management decisions on scientific evidence, but what happens when the scientists disagree?"

You ask why ICCAT, a *tuna* commission, is managing sharks. John explains, "About 350,000 tons of sharks are landed each year, which is about 0.5% of the world's fishery products (Walker 1998)." (You try not to laugh at his quirky habit of citing papers during conversations.) "Many of these species are pelagic and highly migratory, like tunas, placing their management beyond the responsibility of individual countries (Stevens et al. 2000). A substantial portion of shark landings occurs as bycatch in pelagic longline fisheries, such as the tuna fishery. Since ICCAT was already collecting data and managing tuna globally, it made sense for them to monitor and manage sharks in the same way."

Before getting into the specific management details, John gives you some background information on shark biology. "Shark species can be difficult to manage sustainably. Sharks generally grow slower, mature later, and have lower fecundity than bony fishes (Stevens et al. 2000). All sharks have internal fertilization and are either oviparous (egg-laying) or viviparous (live-bearing). Sharks invest a large amount of energy in each offspring, producing only a fraction of the eggs of most teleost fishes. Species with these characteristics may be more susceptible to overfishing and stock collapse.

"Sharks have been caught as bycatch in longline and other fisheries for decades, but it is only relatively recently that some shark species have been targets of an intense directed fishery (Walker 1998). Shark fins are considered a delicacy in some parts of the world, and the demand for fins has resulted in shark fins becoming one of the most valuable food items in the world, selling for more than $400 per fin (Fong and Anderson 2002). This increase in fishing pressure has impacted some stocks, and most fishery scientists agree that many shark populations have decreased in abundance over the past few decades. However, there is debate as to the extent of this decline. Sharks have not been closely monitored over time to identify population trends, so there is a lot of variability in the data and speculation about what the data show.

"And that's why I don't know how to deal with *those*!" he says, pointing to the two papers he had thrown on your desk. You leaf through the documents: Baum et al.'s (2003) article "Collapse and conservation of shark populations in the northwest Atlantic" and Burgess et al.'s (2005) response entitled "Is the collapse of shark populations in the northwest Atlantic Ocean and Gulf of Mexico real?"

You glance up, and John continues, "These two articles demonstrate this disagreement. Both papers were written by respected scientists and published in reputable peer-reviewed journals, but the papers reached different conclusions about the status of shark populations. Baum et al. (2003) estimated that abundance of all recorded shark species, except makos, had declined by more

FIGURE 15.1. Sharks have been caught as bycatch in longline and other fisheries for decades. Recently some shark species have also been targets of a directed fishery for their fins. Above, from left to right: Shortfin mako shark, Porbeagle shark, Blue shark, Spiny dogfish.

than 50% in the past 8 to 15 years. Hammerhead sharks showed the greatest decline in abundance, with an 89% decrease from 1986–2000 (Baum et al. 2003). The authors warned that several species may be at risk of large-scale extirpation (Baum et al. 2003).

“However, Burgess et al. (2005) claimed that these conclusions were overstated, particularly when viewed in the context of uncertain, variable datasets. The authors critiqued Baum et al.’s (2003) choice of data sources and analysis techniques and disagreed with the magnitude of the declines they reported. Burgess et al. (2005) strongly disagreed with the predictions of extinction or large-scale extirpation.

“If we believe Baum’s article, drastic conservation methods would be needed. If we believe Burgess’s article, moderate harvest regulations may be sufficient to allow shark populations to recover. But there is so much uncertainty in the data, it’s hard to know what’s really going on.

“So I need your help,” John continues. “Please read over these two papers. Do you believe Baum, Burgess, neither, or both? Why? What do you think ICCAT should do, regarding shark management? How would you justify this decision to the stakeholders? How does the high level of uncertainty in the data affect your view and management recommendations? Please write your thoughts on these questions, and limit your responses to one page. Thanks so much for your help!”

References

Baum, J. K., R. A. Myers, D. G. Kehler, B. Worm, S. J. Harley, and P. A. Doherty. 2003. Collapse and conservation of shark populations in the northwest Atlantic. Science 299:389–392.

Burgess, G. H., L. R. Beerkircher, G. M. Cailliet, J. K. Carlson, E. Cortes, K. J. Goldman, R. D. Grubbs, J. A. Musick, M. K. Musyl, and C. A. Simpfendorfer. 2005. Is the collapse of shark populations in the Northwest Atlantic Ocean and Gulf of Mexico real? Fisheries 30:19–26.

Fong, Q. S. W., and J. L. Anderson. 2002. International shark fin markets and shark management: an integrated market preference-cohort analysis of the blacktip shark (*Carcharhinus limbatus*). Ecological Economics 40:117–130.

Stevens, J. D., R. Bonfil, N. K. Dulvy, and P. A. Walker. 2000. The effects of fishing on sharks, rays, and chimaeras (chondrichthyans), and the implications for marine ecosystems. ICES Journal of Marine Science 57:476–494.

Walker, T. I. 1998. Can shark resources be harvested sustainably? A question revisited with a review of shark fisheries. Marine and Freshwater Research 49:553–572.

NOTES

NOTES

Case 16

Sampling Gear Biases: Size Structure of Bluegills Collected from the Same Population with Different Gears

The Problem

Students (and biologists for that matter!) have a tendency to accept sampling data at face value. If a gear type primarily collects small fish, then they assume the population is dominated by small fish. If a gear type captures big fish and lots of them, then they assume the population is dominated by large fish. In reality, many biases are possible and actually are very common. To truly understand sampling data, biologists must first understand the biases associated with each gear, and only then will the true nature of the population's structure (e.g., size or age structure) and dynamics (i.e., recruitment, growth, and mortality) be revealed.

Various sampling gears may be differentially effective for different species, and even differentially effective for different sizes of the same species. For example, largemouth bass are commonly sampled with electrofishing gear. The numbers and sizes of largemouth bass collected can vary widely across seasons. During the spring and fall, more and larger largemouth bass tend to be nearshore and vulnerable to the electrofishing gear, which is used in that shallow-water habitat. During midsummer, fewer largemouth bass would be sampled at the same locations because many of the larger bass will have moved offshore to deeper water as a result of the warm summer water temperatures.

In this "gear bias" case study, we will explore the differential size structure of bluegills captured by two common sampling gears, electrofishing and trap nets (also known as modified fyke nets). The trap nets had 1.2- X 1.5-m frames, dual throats, and 19-mm bar mesh. Night electrofishing was undertaken with pulsed DC electricity at approximately 250 V and 8 A. Samples were collected in late May at a water temperature of 23°C.

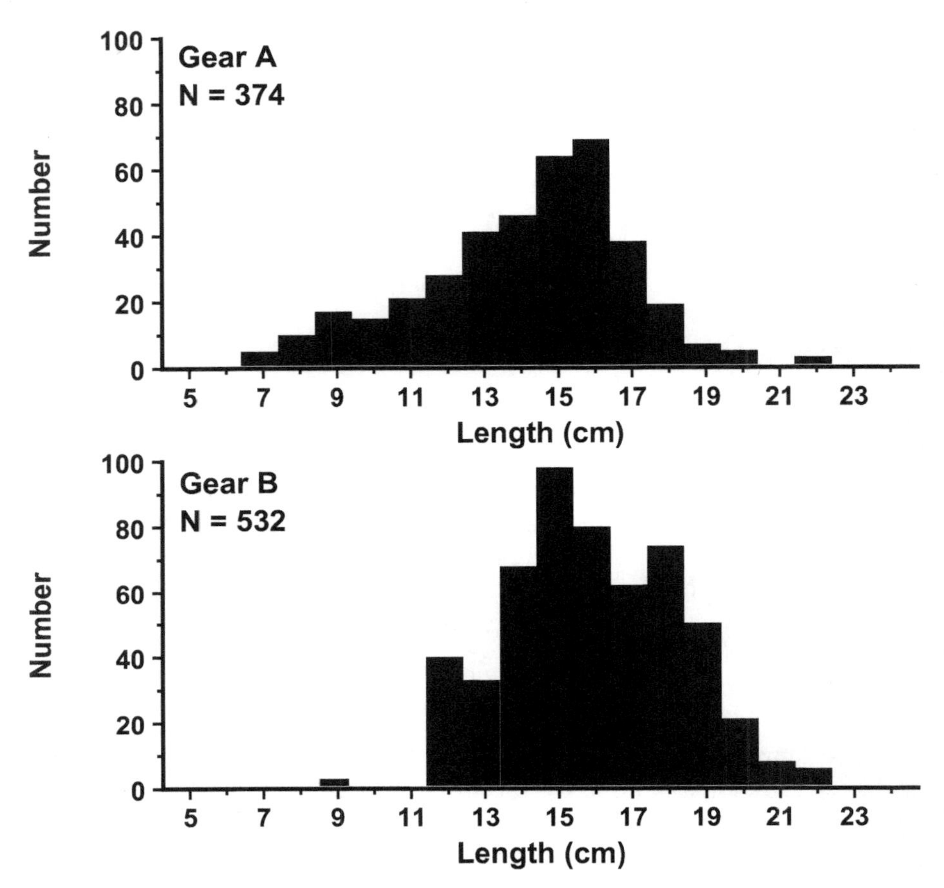

FIGURE 16.1. Length-frequency histograms depicting the size structure of bluegills collected concurrently from Lake Louise, South Dakota, using two gears. One gear was a trap (i.e., modified fyke) net with 1.2- X 1.5-m frames, dual throats, and 19-mm bar mesh. The other sampling method was night electrofishing with pulsed DC electricity at approximately 250 V and 8 A. Samples were collected in late May at a water temperature of 23°C.

Reading Assignment

Prior to undertaking this case study, some background information must be assimilated. If you are not familiar with these two sampling gears, good overviews can be obtained from Hubert (1996) for trap nets and from Reynolds (1996) for electrofishing. Once you have a general understanding of these two gears, then we would like you to read two specific papers—Laarman and Ryckman (1982) and Reynolds and Simpson (1978). Both of these papers deal specifically with the length-related biases for bluegill sampling. As you read them, concentrate on this bias. In addition, we would like you to primarily consider 8-cm and longer bluegills as you read. Bluegills smaller than 8 cm typically are not reliably collected with these two sampling gears.

Questions

Once you have completed your background reading, consider the following information (Figure 16.1). The bluegill population in one 45-ha impoundment was concurrently sampled with both trap nets and by night electrofishing. Sample sizes ranged from 374 to 532 fish, so these data should provide a reliable (precise) assessment of size structure.

1. Does Gear A represent a trap-net sample or an electrofishing sample?
2. Does Gear B reflect the use of trap nets or electrofishing?
3. Explain how you were able to identify the two gears.

References

Hubert, W. A. 1996. Passive capture techniques. Pages 157–181 *in* B. R. Murphy and D. W. Willis, editors. Fisheries techniques, 2nd edition. American Fisheries Society, Bethesda, Maryland.

Laarman, P. W., and J. R. Ryckman. 1982. Relative size selectivity of trap nets for eight species of fish. North American Journal of Fisheries Management 2:33–37.

Reynolds, J. B. 1996. Electrofishing. Pages 221–253 *in* B. R. Murphy and D. W. Willis, editors. Fisheries techniques, 2nd edition. American Fisheries Society, Bethesda, Maryland.

Reynolds, J. B., and D. E. Simpson. 1978. Evaluation of fish sampling methods and rotenone census. Pages 11–25 *in* G. D. Novinger and J. G. Dillard, editors. New approaches to the management of small impoundments. North Central Division, American Fisheries Society, Special Publication 5, Bethesda, Maryland.

NOTES

NOTES

Case 17

Managing Lake Oahe Walleye in the Face of an Imbalanced Food Web

Background

Fisheries biologists are often required to make rapid management decisions during a crisis. Because swift action is usually demanded by the public, biologists rarely have the luxury of taking a long-term, conservative approach to making decisions, but rather rely on prior knowledge to choose a management plan and hope that they predict correctly. This case study will provide you with a real example of fisheries biologists who were faced with choosing a course of action during a rapidly developing situation. Fisheries biologists in South Dakota who manage Lake Oahe (a large Missouri River reservoir) were challenged to save a very large and popular walleye fishery after the prey base crashed.

Lake Oahe is a large impoundment of the Missouri River. It has a surface area of approximately 371,000 ac, has 2,250 miles of shoreline, and is 231 miles long. The walleye fishery in Lake Oahe developed during the 1970s, and was rated among the top 10 walleye lakes in the country from the mid-1980s until 1998 (the year that the prey base crashed). Walleye are supported by a non-native pelagic prey source (rainbow smelt), which became established in the system following escapement from upstream systems (where they were initially brought in as a prey source). When rainbow smelt were abundant, walleye growth rates were very high, reaching 20 inches by age 5. Following a drought during the late 1980s (where water levels and volume declined substantially), the Missouri River drainage experienced a series of wet years during the early-mid 1990s. These conditions enabled Lake Oahe to quickly recharge, and productivity increased dramatically resulting in abundant populations (much higher than levels that had ever been observed in this system) of both walleye and rainbow smelt. So, the mid 1990s were the "boom" years for anglers and biologists on Lake Oahe—anglers were very happy with the fishery, the local economy benefited, and the biologists had achieved a rare benchmark in fisheries science—producing LOTS of BIG fish... However, a compounding series of conditions resulted in the crash of rainbow smelt, starting in 1997.

After refilling during the early 1990s, the high, stable reservoir levels of the mid-1990s were not favorable for rainbow smelt recruitment. Rainbow smelt recruitment is best when reservoir levels slowly rise during the spring spawning period. As such, even though there was a high abundance of rainbow smelt available in 1997, they were primarily composed of older individuals, with few individuals annually recruited into the populations (Figure 17.1). So, the rainbow smelt population in Lake Oahe, which was already decreasing due to a lack of recruitment, was subjected to unusual conditions in 1997 that resulted in an abrupt population crash. This year was a particularly "wet" year for the Missouri River basin and as a result, flood conditions persisted for much of the spring and summer period and discharge through the dams was much above normal (Figure 17.2). This in itself was not detrimental for rainbow smelt, but one more limnological condition occurred that year to complete the chain of events that resulted in the prey crash.

Rainbow smelt are a cold-water species and in a large cool-water reservoir like Lake Oahe, they typically seek the colder waters near and below the thermocline (layer of water with rapidly decreasing temperatures between the warm upper layer—epilimnion, and the cold deepest layer—hypolimnion) during the summer. During 1997 the thermocline became established at similar depths as the outlet tubes for the reservoir and many rainbow smelt entrained in the reservoir discharge (i.e., passed through the dam), which

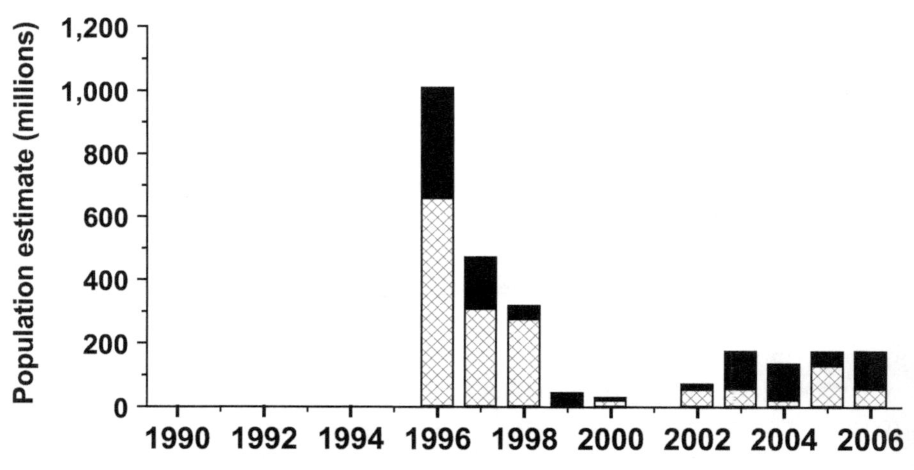

FIGURE 17.1. Rainbow Smelt abundance as estimated by hydroacoustic surveys. Solid bars represent age-0 rainbow smelt abundance, and hatched bars represent age-1 and older rainbow smelt. Note that rainbow smelt abundance was already declining by 1997, as a result of reduced recruitment.

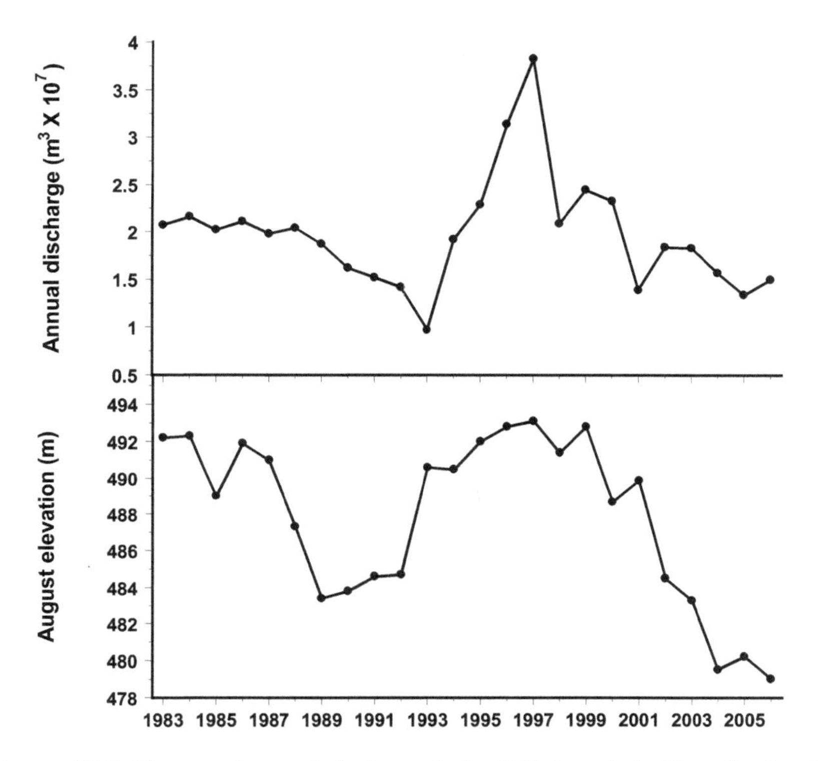

FIGURE **17.2.** Mean total annual discharge ($m^3 \times 10^7$) from Oahe Dam (top) and mean August surface elevation (m above mean sea level) for Lake Oahe, South Dakota.

was very high. Fortunately, there happened to be a graduate student studying entrainment during this year and he estimated that up to 60% of the smelt biomass in Lake Oahe was entrained. Thus, during the course of one year the prey base, which was already in trouble from a lack of recruitment, crashed... and the walleye population suddenly lost their prey base.

The effect on the walleye population was immediately recognized. Walleye abundance, growth rate, and condition declined by 1998, and angler catch rates increased sharply ("hungry" walleye are more likely to be caught by anglers). By 1999 as the condition and growth of walleye continued to decline and rainbow smelt abundance remained low, the fisheries biologists for Lake Oahe were faced with a tough decision about what to do.

The biologists were working from the following set of information:

- Predators and prey were unbalanced in Lake Oahe.
  - o Too many predators and too few prey.

- Rainbow smelt, although much reduced in abundance, would likely recover if predation was reduced.
- Anglers did not want to lose the large walleye in the system. After all, the anglers were just coming through a "boom" period when the number of trophy walleyes harvested was at an all-time high.
- It is nearly impossible to predict reservoir discharge and limnological conditions very far into the future. In other words, biologists did not have the ability to predict limnological conditions in Lake Oahe in any given year. Further, even if reservoir conditions were known, biologists had little control over reservoir water management. Lake Oahe is a multi-use reservoir and is managed for barge-traffic, flood control, irrigation, etc., in addition to recreational opportunities. So, biologists had very little predictive capabilities or management authority for this reservoir—they were at the fate of climate and the U.S. Army Corps of Engineers (who manage the reservoir).
- Any management action would be contentious, given the importance of the walleye fishery, and the myriad of stakeholders—local anglers, resort owners, fishing guides, non-resident anglers (at the time, Lake Oahe was a popular destination for anglers from many states), and the local economy that relied on the walleye anglers.

Assignment

1. Predict (from a food web standpoint) what would need to happen for rainbow smelt to recover, walleye growth and condition to improve, and ultimately for the walleye fishery to recover. Consider the following:
   a. What will happen to the size structure of walleye? Remember that there had been a wide range of walleye sizes in the system.
      i. What sizes are important for anglers?
      ii. What sizes are most affected by a lack of rainbow smelt?
   b. How does your scenario affect all stakeholders?
2. Once you predict an appropriate remedy for the system, what regulation(s) would you pursue to achieve your remedy?
   a. Explain, in detail, how your regulation would accomplish your management objective. Remember, you don't have to change the regulations if you do not want to; you just need to justify your response. The regulations in place were for fish harvest limit: four fish total with only one fish over 457 mm.

NOTES

NOTES

Case 18

Sea Lions: a New Kind of Nuisance

Part 1

> "What a beautiful day to be on the water," Mark thought to himself while parking his car at the Newport Beach harbor. He grabbed a cooler and his fishing gear from the trunk and walked toward the dock. He passed a smashed fish pen and stepped over trash from a garbage can that had been knocked over. "Man, this place is a wreck. I'm going to speak with the harbor manager about th..." He froze in his tracks. His beloved yacht should be moored right here, but the spot was empty. As he looked more closely, he realized the horrible truth as he saw the tip of the mast poking out of the water—his yacht had sunk. A dark brown, whiskery face popped out of the water right next to his boat's mast. The sea lion barked at him as it heaved its massive body onto a nearby boat.

Newport Beach, in southern California, is home to about 10,000 vessels, 1200 offshore moorings, and 1,200 residential and commercial piers. And now it is also a reluctant home to a group of California sea lions *Zalophus californianus*.

About 20 sea lions first arrived in Newport Beach in the spring of 2005 and broke into some fish rearing pens in the bay. The pens were a perfect haul-out location for the sea lions, which spend up to 7 hours per day out of the water to stay warm. While warming themselves on land (or, in this case, a dock), male sea lions make an amazing amount of noise as they battle for territory.

Newport area residents decided to dissuade the sea lions from making this their permanent home because of this initial property damage and the constant noise. Unfortunately, this task proved to be much more difficult than the town expected, and an epic struggle began between man and animal.

When the fish pens were repaired and the sea lions could no longer get in them, the sea lions started hauling out on vessels and docks. Male sea lions measure up to 8 feet long and can weigh more than 800 pounds; when a few of these behemoths lay on a boat or fight on a dock, damage is inevitable. The Newport sea lions damaged vessels, docks, and navigation aids and even sank a couple boats.

FIGURE 18.1. A group of California sea lions sunning on a boat.

In another attempt to get the sea lions to leave, the city eliminated sea lions' access to convenient haul-out sites, such as boats and docks. Owners sea lion-proofed their vessels with snow fence, netting, boards, and other physical barriers. The harbor installed rollers on dock railings so sea lions could not get enough traction to climb onto the docks. The sea lions were undeterred and used any available haul-out location.

Residents also stopped directly or indirectly feeding the sea lions, hoping the animals would leave if a major food source disappeared. No one was allowed to dump food, bait, fish, or fish parts overboard. The sea lions proved their skills as natural hunters and did not leave the harbor when the handouts stopped.

Newport Beach residents decided that they needed a more active approach. Harbor patrols and a motion-activated device sprayed sea lions with water to discourage them from hauling out on boats and docks. Although the sea lions did not like the spray, they learned quickly to avoid patrols and the motion-detecting sprayers. They still did not leave the Newport Beach harbor.

Discussion Questions:

- Do you think the residents of Newport Beach should be trying to "persuade" the sea lions to leave? Why or why not?
- What do you think the city's next steps should be?
- Would your answers differ if this was another species causing the damage and disturbance? If so, how?

Sea Lions: a New Kind of Nuisance
Part 2

Other places with sea lion problems have experimented with different deterrents. At Ballard Locks, Washington, sea lions prey heavily upon salmon, possibly hurting the salmon fishery. Biologists tried to remove sea lions from this area by harassing them with underwater firecrackers, noise, predator sounds, predator models, alarms, and rubber bullets. The sea lions learned quickly to ignore these stimuli. Biologists even captured and relocated particularly troublesome individuals, only to find that the sea lions returned quickly to the capture site. Researchers have not found an effective, long-term, non-lethal approach for removing sea lions. Although some methods work initially, the animals learn to ignore or avoid the deterrent, making it less effective over time.

Other harbors have taken a different approach to invasion by sea lions—if you can't beat them, join them. Or, more appropriately, have them join you as a tourist attraction. By installing floating docks to attract sea lions, waterfront areas may see an increase in ecotourism, potentially bringing customers to restaurants and shops. Many of the problems associated with sea lions, such as noise and property damage, still exist in this "wildlife viewing" type of approach, but residents are generally more tolerant of an animal that brings in visitors and money.

FIGURE 18.2. Some coastal communities install floating decks to attract both sea lions and tourists.

For harbors that are not interested in promoting sea lion-based ecotourism, what options are left? The Newport Beach sea lions bark constantly, are sometimes aggressive towards people, cause property damage, lower water quality through fecal pollution, and may damage in-bay fisheries; they are nuisance wildlife species. When other methods to remove sea lions from Newport Bay have failed, are lethal methods appropriate?

Unlike other nuisance species, California sea lions are protected by the Marine Mammal Protection Act (MMPA). The MMPA was written in 1972 to protect marine mammal species believed to be in danger of extinction (Marine Mammal Commission 2001). The MMPA dictates that marine mammal species must not be allowed to fall below their optimum sustainable population levels. If they are depleted, measures should be taken to replenish species abundance. This legislation prohibits harming or killing any marine mammal in U.S. waters, with certain exceptions for Alaska natives and scientific research. Over the past 3 decades, the MMPA has helped many species of whales and other marine mammals increase in abundance and has prevented further decline of other threatened species (Marine Mammal Commission 2001).

FIGURE 18.3. Sea lions bark constantly, are sometimes aggressive towards people, cause property damage, lower water quality through fecal pollution, and may damage in-bay fisheries.

FIGURE 18.4. Attempts to apply plastic fencing to boat decks were often ineffective.

The MMPA was written to protect threatened species, but California sea lions are far from threatened. There are an estimated 237,000 to 244,000 individuals (NMFS 2003), and the population is estimated to be at its optimum sustainable level. Sea lion abundance is increasing about 6% annually (Read and Wade 2000). As sea lions become more abundant, conflicts like the one in Newport Beach harbor will become increasingly common.

Update: Newport Beach, May 2009. In spite of all that Newport residents had done to deter them in the previous year, the sea lions returned to Newport Beach in the spring of 2009. The battle for the harbor continues…

Writing Assignment:

1. Write a letter to the editor of a Newport Beach area newspaper from one of the following viewpoints. In your letter, make sure to identify the character that you chose. Your letter should be about 1 page long.

   Yourself
   Citizen who lives near the harbor
   Citizen who has a boat in the harbor
   Owner of a harbor-view restaurant
   Member of a conservation organization
   Commercial fisherman
   Sea lion biologist
   Local animal rights activist

Write what you think the city's next step should be and why. Think about how your approach might differ if the "nuisance" species was not a marine mammal. Should the Marine Mammal Protection Act be changed? If so, how?

2. Writing as another person from the list, write a 1-paragraph response to your first letter to the editor.

References:

Marine Mammal Commission. 2001. The marine mammal protection act of 1972 as amended November 2001. Marine Mammal Commission, 4340 East-West Highway, Bethesda, MD 20814.

NMFS (National Marine Fisheries Service). 2003. California sea lion (*Zalophus californianus californianus*): U.S. stock. Stock status report available online: http://www.nmfs.noaa.gov/pr/pdfs/sars/po2003slca.pdf (November 2009).

Read, A. J., and P. R. Wade. 2000. Status of marine mammals in the United States. Conservation Biology 14:929–940.

NOTES

NOTES

Case 19

Size-Structure Assessment for Pallid Sturgeon

The Problem

The pallid sturgeon (Figure 19.1) is a rare fish that is endemic to the Missouri River and the lower half of the Mississippi River. Your challenge for this case study will be to assess size-structure data for three pallid sturgeon population samples.

Background

The pallid sturgeon was formally listed as an endangered species in 1990 (U.S. Fish and Wildlife Service 1990), which prompted various recovery and management actions by biologists, researchers, and resource managers at federal, state, and university levels (U.S. Fish and Wildlife Service 1993). These actions included population augmentation, biological monitoring, and detailed life history research.

The study area included in this case study encompasses the Missouri River from Montana through Nebraska. Six mainstem dams were constructed on the the upper Missouri River (Figure 19.2). These dams blocked movements and perhaps migrations of pallid sturgeon, and may have contributed to their population declines.

Length-frequency analysis

See Table 19.1 for the length-frequency information obtained by trammel netting for three pallid sturgeon samples at three locations along the Missouri River. First, plot a length-frequency histogram for each of the three samples. Next, calculate the proportional size distribution (PSD; Guy et al. 2007) value for each sample (at the least, calculate PSD, PSD-P, PSD-M, and PSD-T). If you need the equations to calculate these indices, see Anderson and Neumann (1996) or Neumann et al. (in press). Stock, quality, preferred, memorable, and trophy lengths for pallid sturgeon are 33, 63, 84, 104, and 127 cm (fork length).

Figure 19.1. Pallid sturgeon collected from the Missouri River in North Dakota.

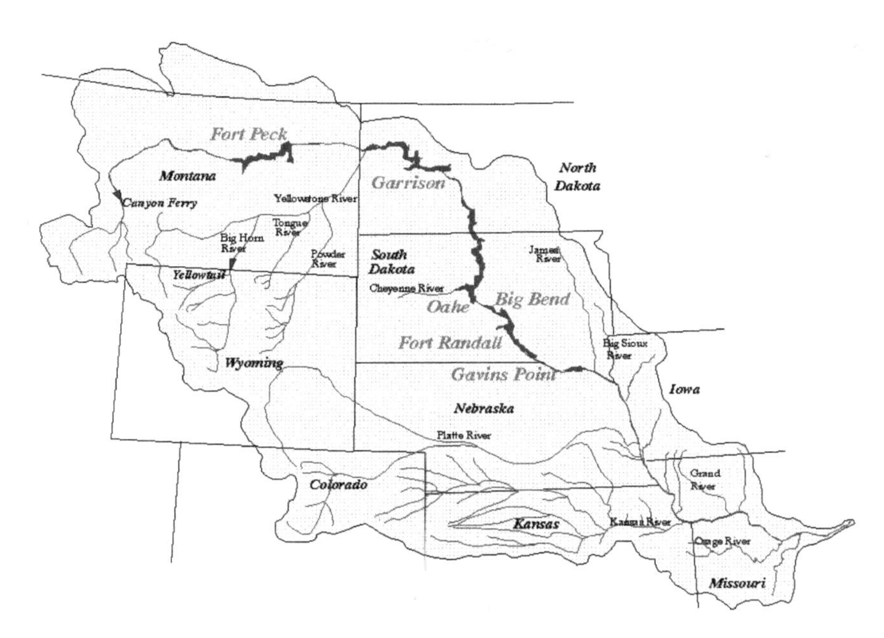

FIGURE 19.2. Missouri River basin, with the six mainstem dams highlighted.

Table 19.1. Number of pallid sturgeon collected with trammel nets and recorded by 3-cm length groups (fork length) from three locations in the Missouri River.

FL (cm)	Location A	Location B	Location C
15	0	0	0
18	0	3	0
21	0	0	0
24	0	7	0
27	0	17	0
30	0	39	0
33	0	38	1
36	0	26	6
39	0	5	5
42	0	7	4
45	0	0	3
48	0	1	1
51	0	1	4
54	0	2	6
57	0	0	7
60	0	3	10
63	0	1	13
66	0	1	11
69	0	0	11
72	0	0	0
75	0	0	1
78	0	0	0
81	0	0	0
84	0	0	0
87	0	0	0
90	0	0	0
93	0	0	0
96	0	0	0
99	1	0	0
102	0	0	0
105	0	0	0
108	2	0	0
111	1	0	0
114	1	1	0
117	2	0	0
120	3	1	0
123	8	0	0
126	5	2	0
129	12	5	0

TABLE 19.1. Continued.

FL (cm)	Location A	Location B	Location C
132	18	4	0
135	24	12	0
138	19	11	0
141	20	17	0
144	18	8	1
147	19	4	0
150	11	3	0
153	12	3	0
156	4	1	0
159	1	1	0
162	4	0	0
Sum	185	224	84

Questions

Based on length frequency and your calculated size structure indices, interpret the status of the three pallid sturgeon populations that were sampled. What can you discern about likely recruitment patterns for the three population samples? Are differences apparent among the three populations? How do apparent recruitment patterns mesh with the concept that this is a federally endangered species?

References

Anderson, R. O., and R. M. Neumann. 1996. Length, weight, and associated structural indices. Pages 447–481 *in* B. R. Murphy and D. W. Willis, editors. Fisheries techniques, 2nd edition. American Fisheries Society, Bethesda, Maryland.

Guy, C. S., R. M. Neumann, D. W. Willis, and R. O. Anderson. 2007. Proportional size distribution (PSD): a further refinement of population size structure index terminology. Fisheries 32:348.

Neumann, R. M., C. S. Guy, and D. W. Willis. In press. Length, weight, and associated structural indices. *In* A. V. Zale, D. L. Parrish, and T. M. Sutton, editors. Fisheries techniques, 3rd edition. American Fisheries Society, Bethesda, Maryland.

U.S. Fish and Wildlife Service. 1990. Endangered and threatened wildlife and plants: final rule to list the pallid sturgeon as endangered. Federal Register 55:36641.

U.S. Fish and Wildlife Service. 1993. Pallid sturgeon recovery plan. U.S. Fish and Wildlife Service, Bismarck, North Dakota. Available: http://endangered.fws.gov/recovery/index.html#plans (October 2009).

NOTES

NOTES

Case 20

Standardized Sampling: Lake Meredith, Texas

Background Information

Many biologists have to deal with clumped distributions of organisms when designing a sampling program, and fishery biologists are no exception (Brown and Austen 1996). Because we tend to have so much variation in samples among individual net catches or individual electrofishing stations, many fishery management biologists use standardized sampling procedures (Bonar et al. 2009). Under this concept, standardized sampling means that sampling gears are chosen to effectively sample the target species and that the same gears are used, to the extent possible, at the same sites, at the same times of year, and from year to year.

Fishery biologists often hotly debate the proper method for selection of sampling sites. Some biologists prefer to subjectively choose sampling sites that they believe will yield substantial numbers of target organisms, while others believe that sampling sites should be randomly selected. Even among the group of biologists who believe in random site selection, there still is disagreement. Some biologists will randomly choose the sampling sites, and then return to those same sites in subsequent years, while others argue that a new group of random sites should be selected and sampled each year.

Regardless of these disagreements, the purpose of standardized sampling is to minimize variation due to sampling device efficiency and seasonal changes in sampling data. In this way, long-term trends in population structure and dynamics can be more reliably monitored. While a one year increase or decrease in catch per unit effort for a target species may not cause alarm, a continuing trend in that increase or decrease over a 10-year period certainly would be a concern.

Setting

Lake Meredith is an impoundment located on the Canadian River in northern Texas (Figure 20.1). It was constructed in 1965, and principal water uses are municipal and industrial water supply, and recreation. The nearest major metropolitan area is Amarillo, which is 60 km from the lake. The surface area at conservation pool is 6,685 ha, with a mean depth 9.2 m and a maximum depth of 38.8 m (Table 20.1). The shoreline development index is 5.05, and the watershed encompasses 15,587 km^2. Secchi disc transparency typically ranges from 0.6 to 2.4 m, while water conductivity typically is near 2,150 uS/cm. There are seven boat-ramp and nine shore-access sites. Four fishing piers meet Americans with Disabilities Act standards for accessibility.

Fish Asssemblage

Species present and a general abundance categorization (high, medium, and low based on catch per unit effort in Meredith and surrounding reservoirs) for fishes in Lake Meredith include: gizzard shad (medium), common carp (low), inland silverside (high), river carpsucker (low), blue catfish (low), black bullhead (low), channel catfish (medium), flathead catfish (high), white bass (high), green sunfish (medium), warmouth (low), bluegill (medium), longear sunfish (low), smallmouth bass (high), largemouth bass (medium), white crappie (medium), black crappie (low), yellow perch (low), and walleye (high).

Your Assignment

Your task will be to assess the physical description of Lake Meredith, Texas, consider the fish assemblage, and design a standardized sampling program for this water body. Remember, standardized sampling means that sampling gears are chosen to effectively sample all target species and that the same gears are used, to the extent possible, at the same sites, at the same times of year, and from year to year. At the minimum, you need to decide on an appropriate gear or gears, and decide what time of year that each gear will be used. Once you have completed your design, your instructor will show you the actual standardized sampling design used by Texas Parks and Wildlife Department biologists.

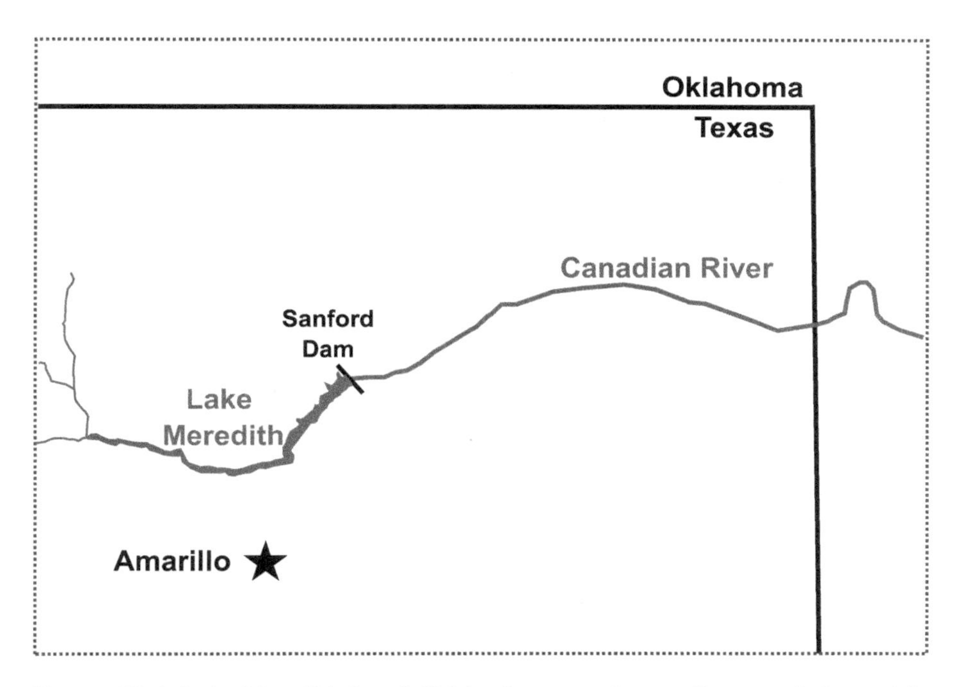

FIGURE 20.1. Lake Meredith is a 6,685-ha (conservation pool) reservoir located in north Texas.

FIGURE 20.2. Dam face (Sanford Dam) at Lake Meredith, Texas.

TABLE 20.1. Habitat survey of the littoral zone and physical habitat types for Meredith Reservoir, Texas. A linear shoreline distance is provided for each habitat type found. "Percent" indicated is percent of total shoreline distance or percent of total reservoir surface area. Habitat surveys were conducted in August when the vegetation growth was at its peak. Eurasian watermilfoil and cattails die during the winter months.

		Shoreline Distance		Surface Area	
Habitat	Type	Km	Percent	Ha	Percent
Water's edge	Concrete	0.3	0.3		
	Rip rap	1.6	1.7		
	Gravel	5.1	5.5		
	Nondescript (mud/silt)	6.1	6.6		
	Flooded terrestrial vegetation	23.9	25.7		
	Boulder	24.5	26.3		
	Rock	31.5	33.8		
Vegetation	Eurasian watermilfoil*			9	0.2
	Cattail			98	2.0
Near shore	Boat docks, piers, marinas	0.5	0.7		
	Dead trees and stumps	0.5	0.7		

* There is probably more Eurasian milfoil present than is recorded on this table. The milfoil extends out from shore and grows in deep water that is not accurately mapped.

FIGURE **20.3.** A steep, rocky point near mid-reservoir of Lake Meredith, Texas.

FIGURE **20.4.** Upper end of Lake Meredith, Texas, during late winter and prior to renewed growth by the emergent vegetation.

References

Bettoli, P. W., and M. J. Maceina. 1996. Sampling with toxicants. Pages 303–333 *in* B. R. Murphy and D. W. Willis, editors. Fisheries techniques, 2nd edition. American Fisheries Society, Bethesda, Maryland.

Bonar, S. A., W. A. Hubert, and D. W. Willis, editors. 2009. Standard methods for sampling North American freshwater fishes. American Fisheries Society, Bethesda, Maryland.

Brown, M. L., and D. J. Austen. 1996. Data management and statistical techniques. Pages 17–62 *in* B. R. Murphy and D. W. Willis, editors. Fisheries techniques, 2nd edition. American Fisheries Society, Bethesda, Maryland.

Hayes, D. B., C. P. Ferreri, and W. W. Taylor. 1996. Active fish capture methods. Pages 193–220 *in* B. R. Murphy and D. W. Willis, editors. Fisheries techniques, 2nd edition. American Fisheries Society, Bethesda, Maryland.

Hubert, W. A. 1996. Passive capture techniques. Pages 157–181 *in* B. R. Murphy and D. W. Willis, editors. Fisheries techniques, 2nd edition. American Fisheries Society, Bethesda, Maryland.

Reynolds, J. B. 1996. Electrofishing. Pages 221–253 *in* B. R. Murphy and D. W. Willis, editors. Fisheries techniques, 2nd edition. American Fisheries Society, Bethesda, Maryland.

NOTES

NOTES

Case 21

Northern Pike Reproduction and Early Life History: Ties to Recruitment Patterns

The Problem

Depending on the fish species, and sometimes even the population, recruitment typically is related to habitat, environmental factors, or both. In this case, you will assess the reproduction and early life history of the northern pike. Then, we will ask you to use your new-found knowledge to devise a future management strategy.

Definition of Recruitment

Recruitment seems to be a straightforward term, but you will find that its definition can be rather slippery. Probably the most accepted definition of recruitment within the fisheries profession is the number of individuals hatched in any year that survive to reproductive maturity. However, alternative definitions might include the number of individuals that reach harvestable size, a particular size or age, or the minimum size captured by a particular sampling gear. In the title of this case study, you could replace "recruitment" with "year-class strength."

Preliminary Library/Internet Assignment

Prior to undertaking this case study, it is essential that you understand northern pike 1) spawning behavior, 2) spawning habitat, 3) behavior of newly hatched fry, and 4) nursery habitat. Information can be found in fish identification books, such as Scott and Crossman (1973) or Eddy and Underhill (1974). Books on fishes from states or regions with northern pike would likely contain similar information. Much of this information might also be found on internet web sites, such as http://www.fishbase.org/search.cfm or http://hatch.cehd.umn.edu/research/fish/fishes/natural_history.html (both sites active September 2009). Finally, your college or university library may provide online literature or database searches with engines such as Fish and Fisheries Worldwide or Google Scholar. Searches on such sites with specific key words can be highly useful to locate scientific articles with the specific information that you seek. Learning how to access such information can be challenging but is an important skill to master.

Setting

A series of six mainstem reservoirs were constructed on the Missouri River in Montana, North Dakota, and South Dakota. Oahe Dam (Figure 21.1), which impounds Lake Oahe in North Dakota and South Dakota, was closed in 1958. The reservoir took approximately 10 years to fill (Figure 21.2). The reservoir has a surface area of 150,100 ha at full pool. Maximum and mean depths are 62.5 m and 18.3 m, respectively. The reservoir is 372 km long, with a shoreline length of 3,620 km and a shoreline development index of 26.4. The trophic status of this impoundment ranges from oligotrophic in the lower end to mesotrophic in the upper end. The lower end of the reservoir thermally stratifies in summer, and the hypolimnion retains dissolved oxygen during this stratification.

Questions:

1. *Initial exercise for the "filling" years:* Based on your newly acquired knowledge of northern pike spawning and nursery habitat, predict the patterns in pike recruitment (year-class strength) for the Lake Oahe population from 1958 to 1968. Remember, these are the years in which the reservoir was filling.

2. *Secondary exercise for subsequent years:* Now, let's turn to recruitment patterns for northern pike after the reservoir reached full pool. What overall trends would you expect to see in northern pike recruitment? Why? Under what environmental conditions would you expect to see strong and weak year classes? Why?

3. *You are a management biologist:* Your current job responsibilities involve northern pike management. In Scenario A, your job in a Midwestern state is to increase recruitment of northern pike in a large reservoir. How might you do so (hint: think of habitat management)? In Scenario B, your job in a western reservoir is to limit natural recruitment of northern pike because of their predation on native salmonids. How could you manage your reservoir to limit northern pike recruitment?

FIGURE 21.1. Oahe Dam, north of Pierre is one of four earthen dams on the Missouri River in South Dakota. The reservoir formed by the dam, Lake Oahe, is 372 km long and has a shoreline length of 3,620 km.

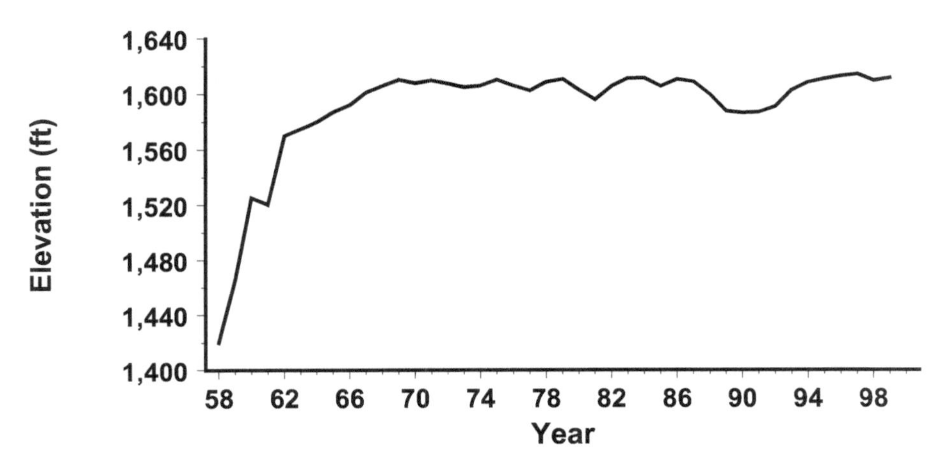

FIGURE 21.2. Elevation (feet above sea level) for Lake Oahe, South Dakota, a main-stem reservoir on the Missouri River. The dam was closed in 1958.

References

Eddy, S., and J. C. Underhill. 1974. Northern fishes, third edition. University of Minnesota Press, Minneapolis.
Scott, W. B., and E. J. Crossman. 1973. Freshwater fishes of Canada. Fisheries Research Board of Canada, Bulletin 184, Ottawa.

NOTES

NOTES

Case 22

Successful Use of a Protected Slot Length Limit to Improve Largemouth Bass Size Structure: But What Happens to the Panfish?

Background

Protected slot length limits (e.g., all fish between 30 and 38 cm must be immediately released) have been used to restructure high-density, slow-growing largemouth bass populations in small impoundments (Dean and Wright 1992; Flickinger et al. 1999; Noble and Jones 1999). When anglers have been willing to harvest the small largemouth bass, the regulation results in improved bass size structure. Typically, the management goal is to reduce density of sub-slot fish, improve growth rates, and increase size structure.

In this case study, you will be asked to think beyond population-level management to the community or assemblage level. As the largemouth bass is a top-level predator in many small impoundments, might changes in bass abundance be related to abundance and resulting size structure of prey species?

The Setting

Watkins Mill Lake (Figure 22.1) is a 42-ha impoundment located 50 km northeast of Kansas City, Missouri. It has a maximum depth of 12.2 m, and secchi transparency measurements typically ranged from 0.6 to 1.2 m at the time of this study. The 770-ha watershed includes pasture, row crops, and forest. The fish community included largemouth bass, bluegills, white crappie, channel catfish, and gizzard shad. The largemouth bass population was managed with a 38-cm minimum length limit (all fish less than 38 cm had to be immediately released) beginning in Year 1 of this study, with the 30- to 38-cm protected slot length limit (Figure 22.2) being imposed prior to the fish community sampling in Year 4.

FIGURE 22.1. Watkins Mill Lake in Missouri.

FIGURE 22.2. A protected slot limit was imposed on largemouth bass in Watkins Mill Lake prior to the Year-4 sampling. The primary objective of this regulation was to improve the size structure of the largemouth bass population.

The Problem

In Table 22.1, we provided the annual sampling data for largemouth bass collected from Watkins Mill Lake by night electrofishing. Was the protected slot length limit effective at improving the largemouth bass size structure? Why?

At the same times and with the same gear, the bluegill population was annually sampled (Table 22.2). In Years 1–3, prior to the imposition of the largemouth bass slot regulation, bluegill proportional size distribution (PSD [Guy et al. 2007]; percent of 8 cm and longer bluegills that also exceeded 15 cm) ranged between 48 and 53. Mean length at time of capture for age-4 bluegills ranged from 165 to 188 mm those three years. Predict what changes, if any, occurred in the bluegill population by Year 9. Write your predictions into the blank cells for bluegill PSD and bluegill mean length at age 4 in Table 22.2.

TABLE 22.1. Size structure and growth of largemouth bass collected by night electrofishing from Watkins Mill Lake, Missouri. A 30- to 38-cm protected slot length limit was imposed prior to the Year-4 sample. N = sample size; PSD = proportional size distribution (the percentage of 20 cm and longer largemouth bass that also exceeded 30 cm); PSD-P = proportional size distribution of preferred-length fish (the percentage of 20 cm and longer bass that also exceeded 38 cm); length at age 3 = mean total length (mm) at time of capture for age-3 fish.

Year	N	PSD	PSD-P	Length at age 3
1	170	a	a	208
2	478	3	1	211
3	299	1	1	241
4	201	42	4	241
5	2,288	25	6	274
6	1,972	41	8	287
7	1,374	32	9	282
8	1,792	59	10	295
9	1,092	50	24	287

a = sample size insufficient to reliably calculate these indices

Table 22.2. Size structure and growth of bluegills collected by night electrofishing from Watkins Mill Lake, Missouri. A 30- to 38-cm protected slot length limit was imposed on largemouth bass prior to the Year-4 sample. N = sample size; PSD = proportional size distribution (the percentage of 8 cm and longer bluegills that also exceeded 15 cm); PSD-P = percentage of 8 cm and longer bluegills that also exceeded 20 cm; length at age = mean total length (mm) at time of capture by age group.

Year	LMB PSD	LMB PSD-P	LMB length at age 3 (mm)	BLG PSD	BLG length at age 4 (mm)
1	a	a	208	48	188
2	3	1	211	49	183
3	1	1	241	53	165
4	42	4	241		
5	25	6	274		
6	41	8	287		
7	32	9	282		
8	59	10	295		
9	50	24	287		

a = sample size insufficient to reliably calculate these indices

References

Dean, J., and G. Wright. 1992. Black bass length limits by design: a graphic approach. North American Journal of Fisheries Management 12:538–547.

Flickinger, S. A., F. J. Bulow, and D. W. Willis. 1999. Small impoundments. Pages 561–587 *in* C. C. Kohler and W. A. Hubert, editors. Inland fisheries management in North America, 2nd edition. American Fisheries Society, Bethesda, Maryland.

Guy, C. S., R. M. Neumann, D. W. Willis, and R. O. Anderson. 2007. Proportional size distribution (PSD): a further refinement of population size structure index terminology. Fisheries 32:348.

Noble, R. L., and T. W. Jones. 1999. Managing fisheries with regulations. Pages 455–477 *in* C. C. Kohler and W. A. Hubert, editors. Inland fisheries management in North America, 2nd edition. American Fisheries Society, Bethesda, Maryland.

NOTES

NOTES

Case 23

Why Does it Look Like That?
How Morphology is Related to Ecology and Management

The Problem

It is only two days into your winter break, and your younger brother is already getting on your nerves. His mission is to make your parents think you haven't been working hard in college. It's part of an elaborate plan that he sees ending with him getting a new game system and with you getting a lecture from your parents.

For today's attack, he walks in holding a nature magazine. Feigning interest, he opens to an article on marine fishes. As soon as your parents walk in the room, your annoying brother says to you, "I know you've been studying fish at college. Why does it look like that?" Your dad naively takes the bait and chimes in, "Yeah, I've always wondered that too. Why is this fish red and that one silver? How about this weird-looking one with the huge teeth—why would it have those? Where would these fish live, and what would they eat?" Your brother just grins, triumphantly.

You glance down at the three pictures they are looking at.

1. Brownish-black, with glow-in-the-dark spots along its belly. Maximum size 25 cm.

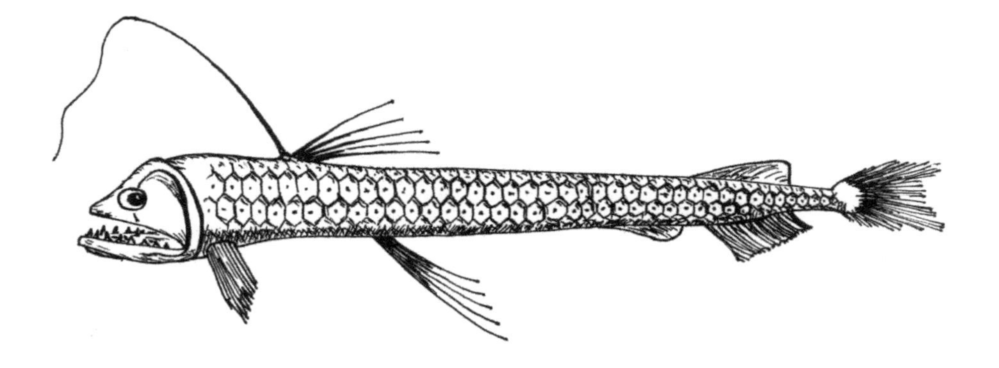

2. Overall very silvery, with bluish-green on its back. Sides and belly are white with dark spots. Two large spots on tail.
   Maximum size 39 cm.

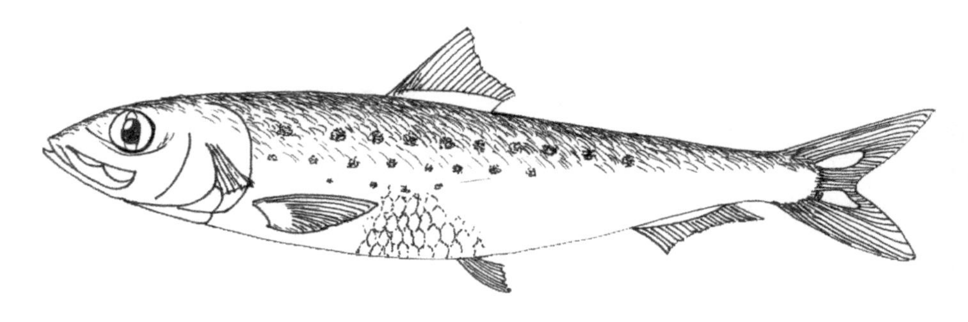

3. Red with small blue spots on top half, pinkish on bottom half. Body is laterally compressed. Maximum size 41 cm.

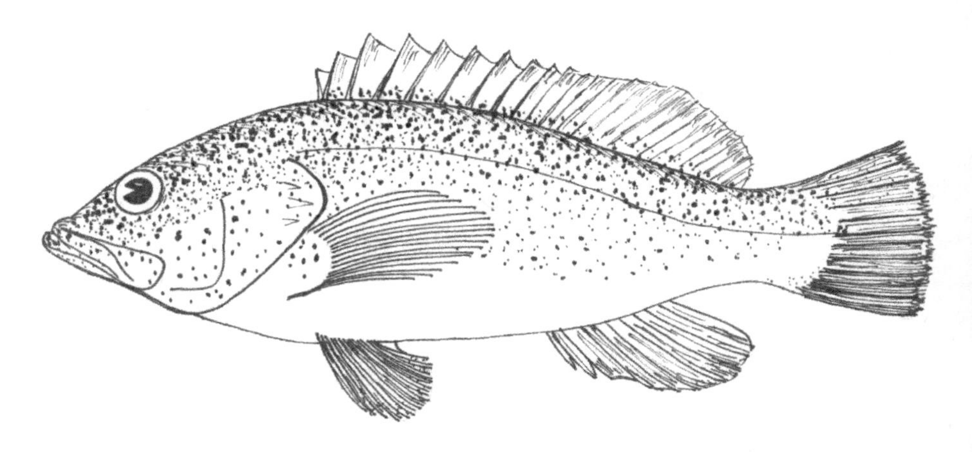

What your bratty little brother doesn't know is that you did, in fact, learn a lot about fishes while at school. While you are not quite sure what species these fish are, you have learned enough about fish morphology, ecology, habitat, and management to tell your family a lot about each fish. Who's getting that new video game system now?!

Your Assignment

Write an explanation for your family about each of these three marine fishes. Based on what they look like, explain where they might live, what they might eat, and any other interesting behaviors (i.e., schooling, attracting prey, etc.) you would predict that they would demonstrate. If you use any specific "sciencey" words (like pelagic, photophores, etc.), be sure to explain them in terms your non-sciencey family could understand. Each fish's description should be about ½–1 page long. Some possible traits to look at: color, body size and shape, mouth and teeth, fin size and placement, special adaptations, etc.

For each trait you describe, explain what it tells you about the fish's habitat, diet, reproduction, or behavior, and why that adaptation is advantageous. For example, a fish that has a mouth on the underside of the head (inferior) and fleshy lips is best suited for bottom-feeding, so you would expect to find this fish at the bottom of a water body. Conversely, a fish with a mouth directly at the front of the body (terminal) feeds on prey in front of it, so it would be more likely to be found in open-water (pelagic) areas.

Based on what you've figured out about the fish, make some educated guesses about its management: do you think it is commercially harvested? If so, what types of gear would work best? Would it be easy or difficult to harvest (and overharvest) this species?

NOTES

NOTES

Case 24

Developing a Pond Management Plan

The Problem

"Hi, this is George Harker. I've got a question about my pond," an elderly voice says through the phone.

Even though you get a lot of these calls each day, your interest is always piqued. Rarely do you get the same question twice, and it's always kind of fun "playing detective" and figuring out what the problem is and how to solve it. "How can I help you, Mr. Harker?" you respond.

"Well, I've got a pond in front of my house. It has some fish in it now, but I'm not really sure what kinds or how many there are. My grandkids are getting old enough to start fishing, and I'd love it if they could catch fish pretty regularly. I don't care as much about getting the huge, trophy fish, and we probably won't keep too much of what we catch. I just want something to keep the grandkids out of trouble when they're visiting. And get them away from that darned TV for a little while." He lets out a gruff laugh and continues, "I think I'd like to have largemouth bass, sunfish, catfish, and rainbow trout in the pond. That's what my neighbor's friend said I should stock. We're not going to let the kids swim in the pond because my wife doesn't want them messing up the house with their wet, muddy footprints. And we don't plan to use the pond for irrigation or livestock. We just want a pond that looks nice where we can catch some fish. What should I do?"

You suggest that you and George meet at the pond tomorrow morning, since you need to get more information about the pond before you make any recommendations. During your visit to the site the next morning, you collect the following information on the pond's construction, water quality, fish, and vegetation:

Construction: The 40-year old, spring-fed pond is round, with a diameter of approximately 65 meters. The pond is 3 meters deep at its deepest point, with a mean depth of 2.2 meters. In the pond, the shoreline banks have a slope of 2.5:1.

Water Quality: You sample the water at a few locations in the pond during the early morning, obtaining these average values:

Dissolved oxygen = 5 mg / L
pH = 6.4
Alkalinity = 18 mg / L
Nitrate = 4 mg / L
Surface temperature = 23° C

Fish: The previous pond owner had stocked largemouth bass *Micropterus salmoides* and bluegill *Lepomis macrochirus*, and George would also like to stock channel catfish *Ictalurus punctatus* and rainbow trout *Oncorhynchus mykiss*, if possible. You and George sample the current population by seining and angling (after he stops envying you for having a job where you actually get paid to go fishing). Here is what you catch:

Seining (4 passes): number of fish in each category of total length (mm)

Species	< 25 mm	25–75 mm	75–125 mm	> 125 mm
Largemouth bass	0	0	1	1
Bluegill	1	8	28	3

Angling (2 people, 60 minutes): total length (mm) and weight (g) of each fish

Species	Length (mm)	Weight (g)	Ws
Largemouth bass	395	1005	108
Bluegill	86	9	82
Bluegill	115	22	76
Bluegill	128	30	73
Bluegill	97	10	61
Bluegill	133	33	71
Bluegill	109	20	83
Bluegill	130	32	74
Bluegill	121	26	76

Fish condition:

One measure of fish condition is relative weight, which is the ratio of a fish's observed weight to the predicted weight of a healthy fish of the same length (i.e., the standard weight). A fat fish will have a high relative weight (>100), while a thin fish will have a low relative weight. A relative weight less than 80 is associated with fish that are very thin and often indicates a lack of food. Relative weights between 80 and 100 are common in healthy populations.

First, calculate standard weight (W_s) using the following equations, based on total length (TL, in mm):

Largemouth bass: $W_s = 10 \wedge (-5.528 + (3.273 \times \log TL))$, and
Bluegill: $W_s = 10 \wedge (-5.374 + (3.316 \times \log TL))$,

The relative condition factor (W_r) is then calculated as observed weight divided by standard weight, multiplied by 100 (Murphy et al. 1991).

Vegetation: You observe the following plants in the pond, and you estimate the percent of the pond area that is covered by each type. George admits to taking a drawing class at the local community college to fill up some empty retirement hours, so you ask him to sketch each type of plant.

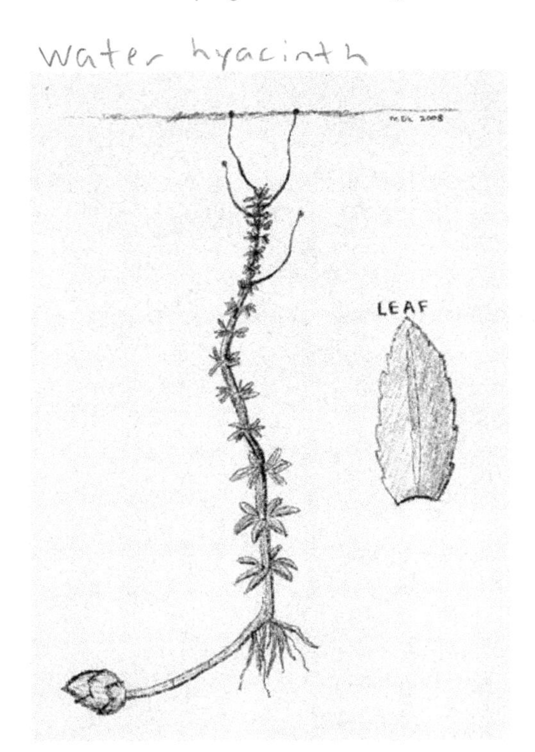

Plant # 1, in about 50% of the pond. This submersed vegetation is found throughout the pond in dense mats. The plants have long stems emerging from a network of rhizomes. Whorls of 3–8 leaves are spaced further apart near the base than at the tip of the plant. Tiny flowers are present at the end of long flower stalks. A kidney bean-sized tuber is present near the base of the plant.

Plant # 2, in about 10% of the pond. This emergent vegetation is growing in clumps and varies in height from just above the surface to about 1.5 m tall. The plants have thick, almost woody stems with leaves rising from near the base of the plant. There are clusters of brownish flowers near the tip of the stem. You observe a number of small bluegill swimming around the base of these plants.

Assignment

Working in groups, research your questions about this pond using the internet, textbooks, and articles. Keep track of references that you use.

Write a management plan for this pond. For each section (Construction, Water Quality, Fish, and Vegetation):

1. Summarize your observations and provide your interpretation.
2. What problems did you identify, if any?
3. Make detailed recommendations on short-term and long-term courses of action, and why you think this approach will be successful.
4. Be sure to consider the pond owner's goals when developing your management approach. If some goals cannot be met, explain why.

Include references or website links to support your statements.

NOTES

NOTES

Case 25

Dear Old Dad: Fisheries Meets the Stock Market

Dear John,

I hope all is going well for you in your last semester at college. I still wish that you had gone into a real field like business or engineering, but I guess we'll find out soon whether you can actually make a living as a fisheries biologist. Anyway, maybe you can put all that book knowledge to work for your dear old Dad, because I actually have a fisheries question for you.

Your uncle George called today with a hot tip on a stock that he says I should buy immediately, because it looks like this company is going to make a killing in the next few years. It's some company that makes that fish oil stuff that health nuts take every day. The fish oil comes from some nasty fish that isn't much good for anything else, so it might as well be turned into a useful product. More and more people are becoming health-food wackos these days, so it certainly seems like this is a business that could grow. George got this idea from some investment newsletter that he pays big bucks for, so I'm sure the information is good. Plus, George knows all about this company and that little fish because they're both around the Chesapeake Bay where he lives. I've attached the stock tip that I got from George so that you can read about it. It sounds like a great investment to me, but I wanted to ask you about it since you're always talking about fish, fisheries, conservation, management, yada yada yada...

Here's your chance to show me that all the money I spent on your education was worth something! Get back to me ASAP (this week) on this, because I want to send George some money quickly to get in on this deal. In fact, I think I'll use the cash that I <u>won't</u> have to spend on your tuition next year!

Later (but not too much!),
Dad

Bill's Tiny Gem

Houston-based **Omega Protein** (NYSE: OME) is a fish oil and fish meal processing company. Its main product is derived from menhaden, an abundant inedible fish that the Menhaden Resource Council (MRC) at www.menhaden.org calls "a remarkable citizen of the sea." It also says that fishing for menhaden is "probably America's oldest business," having been established by Native Americans and taught to the first colonists.

We like old, stodgy businesses here, so the MRC's attempts to give prominence to its dues-paying fish work for us. Omega Protein's products are used for health-food applications — science suggests that the omega-3 fatty acid found in menhaden contains benefits for cardiovascular health, cancer, and arthritis. Who knows, perhaps someday it will help Tom regrow some hair.

Omega Protein's products come in the form of nutraceuticals, as well as in a food additive. The stock has been beaten up as weather patterns in 2004 caused a disruption in the menhaden catch, hammering Omega Protein's operations. I expect that 2005 will be much better for the company, and am pretty sure the menhaden, in keeping with its august history, isn't going anywhere. As such, this stock offers a compelling value. *Buy below $7.50.*

Box 25.1. Stock Tip

Your Response to Dad

Okay, here is your chance to show your dad that, yes indeed, you have learned something useful in your time studying fisheries biology and management.

1. Write a letter back to your dad, telling him whether buying stock in Omega Protein Corporation (OME) is a good idea.
2. Make specific references to what you consider to be the critical issues in the 'stock tip.'
3. Support your recommendation with concrete statements of fact, and a detailed explanation of your critical analysis of the current situation with OME and the natural resource on which their business relies. Is the fish oil business one that you can see being able to grow, as George's stock tip predicts?
4. As a biologist, your major analysis should revolve around biological and ecological issues, but feel free to dig into the business model

and financial health of OME. After all, ***fish are biological entities***, but ***fisheries are processes to extract human benefits from those biological entities***. OME is in the business of extracting benefit (profit) from a renewable natural resource (the menhaden fishery). Is their business model one that can be sustained and grown to produce the investment profits that your dad and uncle are looking for?

Financial Evaluation of Potential Purchase of Company Stock

The financial decision to make a stock purchase is simple, in theory. If I buy this stock now, can I sell it at some point in the future for more than my purchase price? If yes, then buy it; if no, avoid it!

Financial websites contain a wealth of information about individual corporations, including those involved in natural resources exploitation like OME. Here are some suggestions for finding information on OME.

- Log into www.finance.yahoo.com
- Enter "OME" in the search box (top left)
- Click "Get Quotes"
- Read information at "Business Summary" and "News and Info" to learn about the company
- See whether "Analysts" (the people who make recommendations for big investment firms) recommend buying this company
- Check out the financial condition and history of OME under "Financials"

FIGURE 25.1. Atlantic menhaden (*Brevoortia tyrannus*)

NOTES

NOTES

Case 26

Exotic Species, Economic Development, and Native Fish Restoration: Are All Possible?

The Problem

You have been hired as a consultant to advise local scientists and government agencies in Santa Catarina, Brazil on approaches to restore native fishes to previously polluted streams from which they had been eliminated. The government hopes to stimulate a sport fishing-based tourist industry in this economically depressed area. On your initial visit to the area you learn that rainbow trout from North America have already been stocked in many of the streams of interest. Can native fish populations be successfully restored where these trout presently exist? Would native fisheries be capable of supporting extensive sport fisheries? What course(s) of action will you recommend for the restoration program?

The Setting

Brazil is a country of great ecological contrasts. The northern section straddles the equator and is one of the world's hotspots for biodiversity. The more southern areas are sub-tropical and temperate in nature and bear resemblance to areas of North America. The median latitude of the state of Santa Catarina is around 27° S (the mirror image of Florida in the United States), however the altitude makes the climate more similar to the Appalachian areas of North Carolina, Tennessee, and Georgia.

The fish fauna

The freshwater fish fauna in Brazil is dominated by many species of the family Characidae, a large and diversified fauna. Familiar characins include the carnivorous Amazon piranhas and the vegetarian pacus, but hundreds of other species of all morphological and ecological descriptions populate the family as well. Other taxa represented in Brazil include the catfish families Heptapteridae and Pimelodidae.

FIGURE 26.1. Map of Brazil showing the state of Santa Catarina in black.

Biologists in Santa Catarina are interested in possible restoration of several species of stream fishes. Chief among these are the tambica, traira, dourado, sorubim, jundia, and lambari. Tambica are a series of several congeneric characin species that are ecologically similar to northern pike. Traira are another top-predator characin. Dourado, also called golden trout, are a salmon-like characin. Sorubim are several large pimelodid catfishes. Jundiá is a smaller catfish that is the subject of a growing aquaculture industry. Finally, lambari are a shad-like forage species that are popular as restaurant fare, where they are served as a deep-fried appetizer (similar to North American rainbow smelt).

Rainbow trout from Denmark were first introduced to Brazil in 1949, when eyed eggs were stocked in the headwaters of rivers in the Serra da Bocaina region, in the states of São Paulo and Rio de Janeiro (MacCrimmon 1971). Some people felt that the native Brazilian fish fauna did not include any suitable species for developing attractive sport fisheries of the types found in Europe and the USA. Trout were again imported to Brazil from California in the 1960s for stocking in local rivers and to create a culture industry (Porto-Foresti et al. 2000). A number of trout culture stations were developed by

Figure 26.2. Lambari and Tambica.

Figure 26.3. Dourado and Traira.

IBAMA (the federal environmental protection agency) to produce trout for distribution throughout the country. Ironically, such distribution is counter to existing environmental laws in Brazil, although IBAMA has made a policy decision not to enforce this regulation, pending more information on the interactions of trout and native species.

Stream habitat

Water quality parameters of the streams in southern Santa Catarina are similar to lowland streams in the southwestern USA. Strongly seasonal rainfall patterns can lead to very low stream discharge and water levels (< 0.5 m in many places) in the winter. Annual mid-column water temperatures vary 12°–24°C.

Industry and economic development

Southern Santa Catarina historically relied on logging and paper production. Activities in these industries negatively impacted water quality in the local streams, and many fish species disappeared. The main forestry cash crop was the Araucaria pine, an indigenous lumber tree that was severely overharvested due to its high value. Araucaria are now protected by law throughout Brazil, and Santa Catarina's forestry industry has now shifted to plantation culture of several Canadian pine species. The forestry industry is much reduced from former levels, and this reduction coupled with modern environmental regulations has resulted in much-improved stream water quality. Stocking and restoration of native species may now be a viable alternative.

Small towns in Santa Catarina that used to rely on forestry are looking for alternative economic development opportunities. Many are interested in tourism development, and they see viable sport fisheries as a critical component for tourism. In hopes of attracting visitors, some municipalities have already stocked rainbow trout into local waters. Trout fishing guide services already exist in some areas, while other municipalities prohibit all fishing until they feel that trout are well established. No systematic survey of either trout or native species has ever been conducted in most of the streams of Santa Catarina.

Impacts of non-native species

A few biological surveys have been conducted in Santa Catarina as part of the EIS (environmental impact statement) for a series of hydropower development projects on the Rio Uruguai. Biologists who conducted these surveys offer the following about the potential impacts of non-native species on indigenous fishes:

> "Populations of stream fish in southern Brazil are most likely regulated by abiotic factors, primarily stream level and water temperature. Populations are kept low by extreme environmental variability, and it is unlikely that ecological interactions (e.g., competition or predation) with introduced fishes will have noticeable impacts on native fish populations." (C. Bizerril, Dorector, NP Consultoria Ambiental, Rio de Janeiro, Brazil).

The Mission

You have been brought to Santa Catarina by Partners of the Americas (POA), a non-profit organization that attempts to stimulate development in areas of Central and South America through the facilitation of technological, scientific, and cultural exchanges with regions of North America. POA has asked you to work with scientists at UDESC (the State University of Santa Catarina), UFSC (the Federal University of Santa Catarina) and EPAGRI (the Cooperative Research and Extension Service of Santa Catarina) to assess the feasibility of initiating native fish restoration projects in the region. You have been charged with assessing the local ecological and political climate for such projects as well as evaluating the likelihood of such projects resulting in a viable sport fishing industry that will stimulate local tourism development.

Questions

1. Are present water quality conditions suitable for the native species mentioned as targets for restoration in the waters of Santa Catarina? Is there more information that you feel you need to make this judgment? How might such information be collected? On what will you base your decisions if no more information can be obtained?
2. Assuming that restoration is to be pursued, how would you design such a program? Discuss the major steps that you would take to create such a program. How would you later evaluate the success of your program?
3. Is it true that aquatic species populations can be regulated by abiotic factors? What evidence can you find for or against this assumption? What is your educated opinion of the statement that native fishes are unlikely to be impacted by introduced species? Support your contentions.
4. Are any of the native species likely to produce populations that will support what you would consider to be quality sport fisheries? What criteria are you using to judge sport fisheries?
5. What are the likely interactions between the native species and introduced rainbow trout? Are there ecological studies that you would recommend that IBAMA initiate in order to investigate these interactions?
6. Are the trout likely to produce self-sustaining populations in this area? What is the likelihood that introduced trout can support a viable sport fishery? What will be the critical factors controlling the establishment of such a fishery? What management actions would be necessary to make such a fishery a trophy fishery that will attract the attention of sport fishermen? Is this fishery ever likely to receive international notice?
7. Given all that you know about this geographic area, the political climate, and the fish species involved, what will be your recommendations to POA and the local scientists? Should native fish restoration be pursued? Should trout stocking be continued? How should this area be managed to have the best chance of establishing a viable sport fishery?

References

MacCrimmon. H. R. 1971. World distribution of rainbow trout (*Salmo gairdneri*). Journal of the Fisheries Research Board of Canada 28:663–704.

Porto-Foresti, F., C. Oliveira, Y. A. Tabata, M. G. Rigolino, and F. Foresti. 2002. Analysis of

NOTES

NOTES

Case 27

Managing a Small-Scale Trophy Largemouth Bass Fishery for Tourism Development

The Problem

You have been asked to advise a local landowner/developer in western Mexico regarding his wishes to develop a trophy largemouth bass fishery in a small private lake. He plans to build a small hotel on the lake that will cater to high-income tourists who are interested in catching the largest bass of their life. Can this lake support a sufficient number of trophy bass to satisfy his high-end clients? Can the bass fishery create sufficient alternative employment to support the commercial fishers who own half of the lake, and who currently gillnet tilapia from the lake as a supplemental income source? Can the bass fishery be developed and the tilapia fishery be improved at the same time? What course(s) of action would you recommend?

Background

"Would you like to see what I am going to turn into the most beautiful bass lake in Mexico?" asked Carlos Rodriguez. Carlos is definitely the entrepreneurial type. He retired early, but is always hatching plans for new business enterprises. After creating an exclusive private beachside housing development north of Puerto Vallarta, he and his brother recently became partners with a small group of commercial fishers who own a unique spring-fed impoundment in the rugged central Pacific coastal mountain region. "I want to create a fishery that will attract bass anglers from all over Mexico and the United States," Carlos said. "They will come here to catch the biggest bass of their lives, and if they don't catch a bass of at least 4 kilos (10 pounds), then they don't pay. I want you to tell me how to create this bass fishery, and I am willing to pay you well to do it." Intrigued, you agree to accompany Carlos to see the lake the following day.

The Setting

After a 3-hour drive from Guadalajara, first through the dry, tequila-producing region of Jalisco state and then through the winding mountain roads and

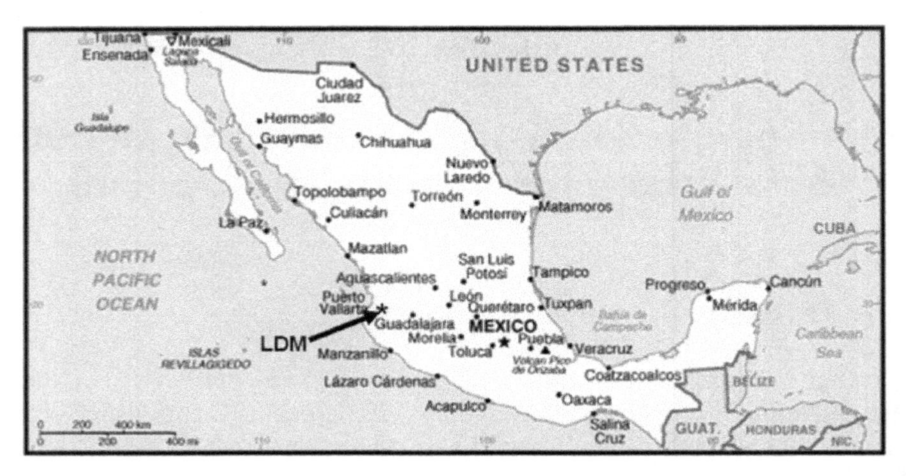

FIGURE 27.1. Map of Mexico, indicating the location of Laguna del Mastranzo (LDM).

extensive volcanic lava fields of southern Nyarit state, you and Carlos arrive at Laguna del Mastranzo (LDM). At only 60 hectares LDM is relatively small. Surrounded by swaying palms and covered with every type of waterbird imaginable, this spring-fed (5700 l/min) impoundment was built in the 1980s by the federal government to create fishery development opportunities for a nearby small town. Submersed aquatic vegetation (of unknown species) is present in at least half of the lake. The lake was initially stocked with various tilapia species of mixed genetic origin. The present fish community includes some native gobies (Gobiidae) and a few other small native stream species, but is dominated by stunted tilapia of mixed genetic origin. Outflow from the lake is directed into a concrete tank (50 m x 25 m x 2 m) that currently contains tilapia fingerlings being raised for commercial sale.

Resource Users

Waiting for you at the lake is a group of a dozen or so commercial fishers who own the lake as 50:50 partners with the Rodriguez brothers. These fishers harvest tilapia from the lake and wholesale them to local fish markets. The fishers demonstrate their cast-netting technique for you, capturing approximately 25 tilapia (100–200 mm TL) with a single cast of the net. With another smaller-mesh net they also capture a few samples of the small native fish species, and some tilapia as small as 70 mm. Your discussions with the fishers reveal that they are disappointed in the average size of the tilapia because they are below desirable market size, and in recent years they have reduced their harvest to a single week each year. During that week they harvest approximately 7,000 kg of fish using gill nets and cast nets. They express

FIGURE 27.2. Laguna del Mastranzo, Nyarit, Mexico.

a ready willingness to eliminate their tilapia harvest if the lake can be converted to another fishery that will provide better income opportunities.

Development Plans

After the fishing demonstration, Carlos explains the plan that he has for developing the lake as a tourist destination. He plans to build a small (10- to 20- room) hotel on the lake to cater to high-end recreational anglers. The hotel will provide package services with fishing guides (the present commercial fishers), boats and motors, refreshments, etc. Carlos believes that the present tilapia population will provide abundant prey to produce trophy-sized largemouth bass (4-6 kg), although he is willing to stock additional prey (he says he has a source for fingerling bluegill). To establish the largemouth bass population, Carlos plans to bring a large number of adult fish from another lake. "I want to bring as many fish as possible to this lake, and I want to bring the biggest fish possible so that the trophy fishery will become established immediately. I want to have 14 or 16 anglers on the lake every day, 2 anglers to a boat. And I want every angler to catch at least one bass >4 kg during their 2-day stay at the hotel," Carlos tells you. "Tell me how to do this, and tell me whether the tilapia population can be simultaneously exploited to support the bass population and to produce commercially acceptable tilapia."

For long-term viability of the population, Carlos tells you that he is willing to stock bass annually, although he is curious about "experiments with mono-sex largemouth bass to produce all female fish with more trophy potential."

Laguna del Mastranzo
Water Quality Data

Surface temperature: 25–33°C (annual range)
Dissolved oxygen (measured in January): 8.4 ppm surface to bottom
Water conductivity: 300 *u*S/cm (sampled in January)

Questions

1. Are tilapia a suitable forage species for largemouth bass? Do they produce enough prey of the appropriate size at the right time of year? Is their mouth-brooding behavior a problem? Can other prey species be used in conjunction with tilapia? Could tilapia be removed completely to start with another forage? Can you speculate on the probable standing crop of tilapia at present?

2. What can be done with the concrete tank? Raise largemouth bass? Raise tilapia?

3. Describe the probable reactions of the tilapia population to bass introduction. What will happen to the population structure of the tilapia?

4. Can tilapia be commercially harvested from a lake that is to support a trophy largemouth bass population? What cautions can you offer?

5. What subspecies of largemouth bass should Carlos stock, and why?
floridanus

6. What is the likely productivity of this lake for largemouth bass? What level of standing crop (kgs/ha) of largemouth bass might you expect?

7. What are the implications of the submersed aquatic vegetation in the lake? How might it affect predation of largemouth bass on either tilapia or bluegill? How would it affect subsequent length frequencies of all three species?

8. What type of largemouth bass population structure would you target to produce a large number of trophy fish? How would you accomplish this management objective? How many trophy fish could you expect to have in this lake at any given time?

9. How frequently can Carlos' trophy largemouth bass be caught? How might the largemouth bass be affected by frequent catch and release?

10. What equipment and techniques will be necessary to move adult largemouth bass from one lake to another?

11. Will it be possible to continue commercial harvest of tilapia in the presence of largemouth bass? What interactions can you predict between these two fisheries?

12. How might the presence of tilapia affect largemouth bass reproduction in this system?

NOTES

NOTES

Case 28

Managing Prey Resources in Colorado Reservoirs

Background

Many large western reservoirs are managed with non-native fish assemblages. In higher elevation reservoirs (cold water systems) a popular sport fish assemblage created by managers was lake trout (top level predator) and kokanee salmon (lacustrine sockeye salmon) and/or rainbow trout (mid-level planktivores and insectivores). This simple fish assemblage is easy to manage (fish culture and stocking techniques are well known) and is very popular with anglers. After the establishment of many reservoir fisheries with this combination, managers sought to increase kokanee and rainbow trout growth rates by introducing an additional invertebrate prey, opossum shrimp. These freshwater shrimp are native to the great lakes region and parts of Canada, and were first introduced into western U.S. systems in 1949 (Kootenay Lake, British Columbia; Spencer et al. 1991). This initial introduction resulted in marked increase in growth of kokanee, prompting managers to introduce opossum shrimp into many other systems, including several Colorado reservoirs. However, most of these introductions had unintended consequences...

The focus of this case study will be Lake Granby, a high elevation reservoir in northern Colorado. Opossum shrimp were first introduced into Lake Granby in 1971, and several years of research and monitoring are available to determine the effectiveness of this stocking. Your assignment is to predict possible food web effects of introducing opossum shrimp into Lake Granby.

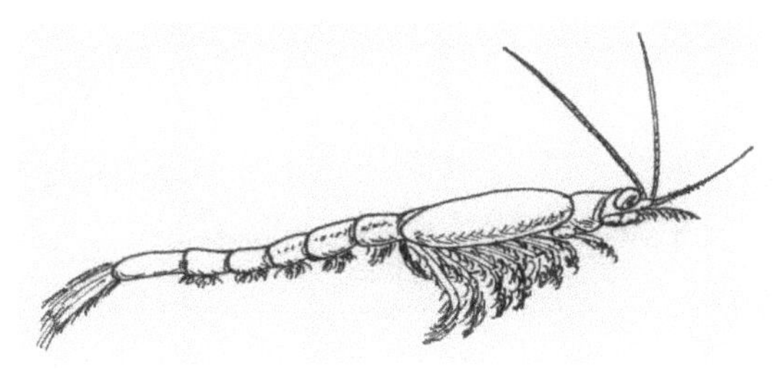

FIGURE 28.1. Opossum shrimp (*Mysis relicta*)

Below you will find the necessary background information to predict the effect of introducing opossum shrimp into Lake Granby.

Lake Granby is a 2,939 hectare, high-elevation (2,523 m) reservoir with a maximum depth of 67 m, and an average depth of 23 m. In other words, this is a large, deep system. The Lake Granby food web is depicted on the next page.

It is your task to insert opossum shrimp into the food web provided. You can assume the following:

- Lake trout are piscivorous (primarily fish diet) at larger sizes and consume invertebrates at smaller sizes (e.g., <500mm total length). Lake trout can reside in the lower water column and readily change depths to forage.
- Kokanee consume zooplankton throughout their life. Kokanee tend to reside in the upper and middle water column.
- Zooplankton consume phytoplankton and primarily occur in the upper water column.
- Opossum shrimp consume zooplankton, but also undergo a diel migration wherein they feed in the upper water column at night, but migrate to deeper water during the day.

Once you have delineated the Lake Granby food web, make predictions about the direct and indirect effects of opossum shrimp. Be sure to discuss all trophic levels (zooplankton, kokanee, and lake trout).

Note: opossum shrimp can directly influence trophic levels via predation or energy flow (being consumed), and if these direct effects are strong, they can also indirectly influence other trophic levels.

References

Johnson, B. M., P. J. Martinez, and J. D. Stockwell. 2002. Tracking trophic interactions in coldwater reservoirs using naturally occurring stable isotopes. Transactions of the American Fisheries Society 131:1–13.

Martinez, P. J., and W. J. Wiltzius. 1995. Some factors affecting a hatchery-sustained kokanee population in a fluctuating Colorado reservoir. North American Journal of Fisheries Management 15:220–228.

Spencer, C. N., B. R. McClelland, and J. A. Stanford. 1991. Shrimp stocking, salmon collapse, and eagle displacement. BioScience 41:14–21.

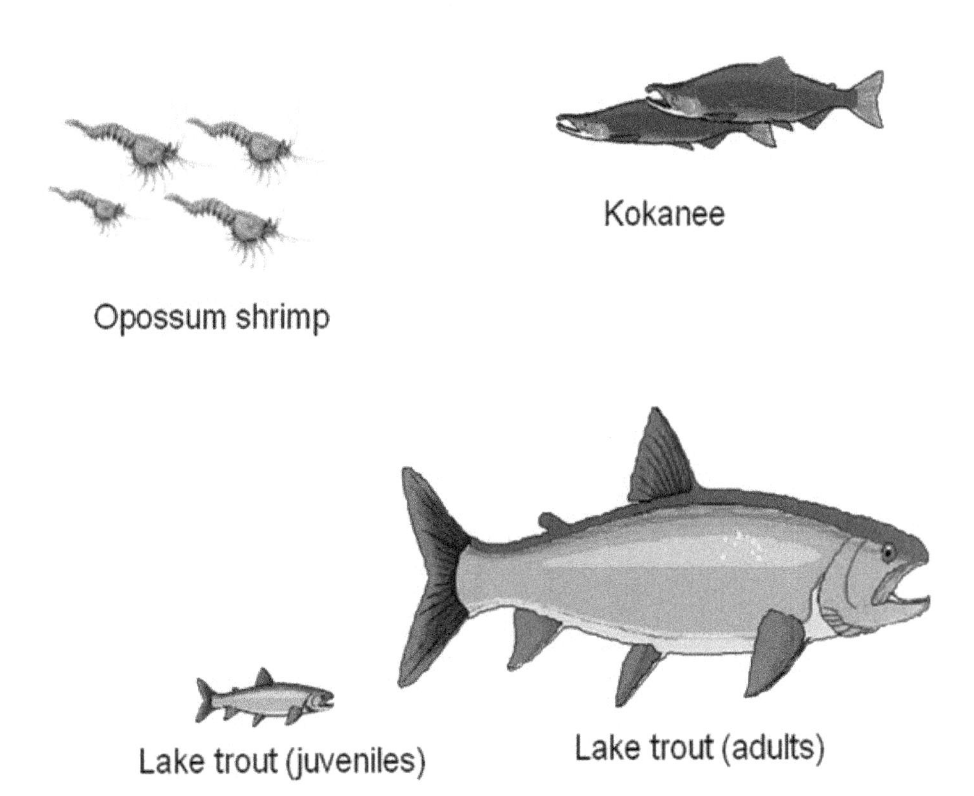

FIGURE 28.2. Lake Granby food web. You must insert appropriate arrows to denote energy flow, and then predict (via arrows) where opossum shrimp will integrate into the food web.

Notes

Notes

Case 29

Can We Make Overfishing Less Prevalent around the World?

Background

The *Code of Conduct for Responsible Fisheries* (FAO 1995) is a voluntary international consensus regarding the most important efforts needed to control overfishing, developed by the Food and Agriculture Organization (FAO) of the United Nations. The Code outlines detailed standards for the scientific, sustainable, responsible and equitable exploitation of fishery resources. Measuring against the criteria in the Code, a recent evaluation by Pitcher et al. (2009) found that only a few nations receive 'barely passing' scores for their efforts to sustainably manage fisheries under their jurisdiction, and most nations get 'failing' scores.

Assignment

Review the FAO Code of Conduct for Responsible Fisheries, and the analysis by Pitcher et al. (2009). Consider the questions on the next page.

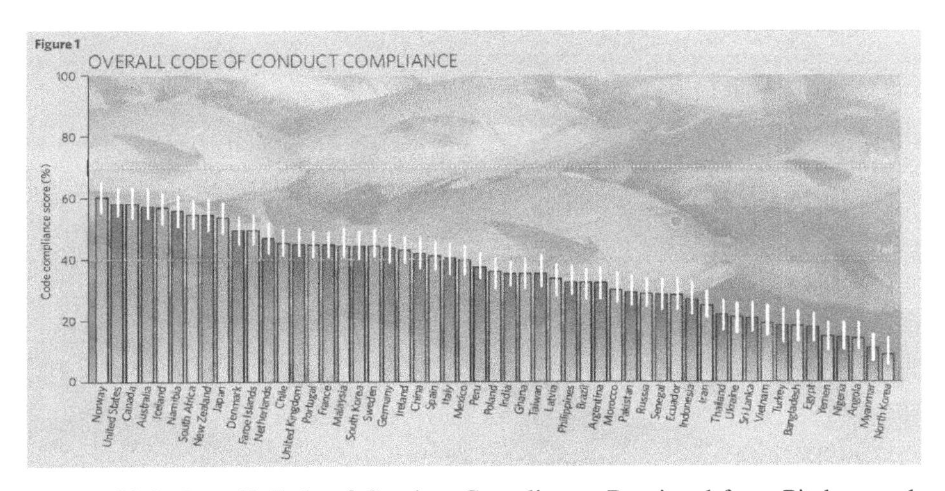

FIGURE **29.1.** Overall Code of Conduct Compliance. Reprinted from Pitcher et al. (2009) with permission from Macmillan Publishers Ltd.

Questions

1. What are the major factors that lead to overfishing? Why do nations allow overfishing to occur in their waters? Could ending overfishing increase ecological, economic, or social benefits for individual countries?
2. Are the criteria in the Code reasonable measures of a nation's efforts toward fishery sustainability? Why or why not? Are there other important factors that should be considered?
3. What patterns are evident in the results of the analysis by Pitcher et al. (2009)? Can you think of other factors besides those discussed by the authors that might explain the results of their evaluation?
4. What factors seem to be the most important in influencing whether a country is successful (or making a significant effort) in addressing overfishing in their territorial waters? Why are these factors the most important?
5. What global benefits would accrue if national scores on the 'overfishing report card' (Pitcher et al. 2009) were to be improved? What mechanisms would increase these benefits, and who would benefit most?
6. What global, political, and economic factors will be most important to change if nations are to do better in their efforts to meet the Code?
7. Many would argue that overfishing, even in sovereign territorial waters, influences the global ecological balance in the oceans as a whole, and thus impacts all peoples on our planet (Alder et al. 2008; Heithaus et al. 2008). Should nations somehow be forced to do more to address overfishing? What can be done to either help or force nations to do better in addressing overfishing?

References

Alder, J., B. Campbell, V. Karpouzi, K. Kaschner, and D. Pauly. 2008. Forage fish: from ecosystems to markets. Annual Review of Environment and Resources 33:153–166.

FAO 1995. Code of Conduct for Responsible Fisheries. Available: ftp://ftp.fao.org/docrep/fao/005/v9878e/v9878e00.pdf (February 2010).

Heithaus, M. R., A. Frid, A. J. Wirsing, and B. Worm. 2008. Predicting ecological consequences of marine top predator declines. Trends in Ecology & Evolution 23:202–210.

Pitcher, T., G. Pramod, D. Kalikoski, and K. Short. 2009. Not honouring the code. Nature 457:658–659.

Notes

Notes

Case 30

Should We Eliminate Catch-and-Release Angling?

Background

Several studies have attempted to determine whether or not fish feel pain. Proponents of this idea have argued that recreational angling (and catch-and-release angling in particular) causes pain and suffering in fish and should be eliminated. Opponents argue that recreational angling is justified because fish do not experience pain and suffering, at least not in a similar way as humans. Moreover, they argue that the definition of pain and suffering is anthropomorphic, and point out that fish lack a neocortex region in their brain and may not have the ability to experience pain. Because pain and suffering are difficult to quantify, or even define, Cooke and Sneddon (2007) attempt to change the focus of this debate from whether or not angling causes pain and suffering to how angling affects the "welfare" of fish. This approach focuses on the quantifiable science of the effects of angling and the growing body of literature available on this topic.

Assignment

Review *Animal Welfare Perspectives on Recreational Angling* by Cooke and Sneddon (2007). Consider the following questions:

1. Define fish welfare. How does this definition differ from pain and suffering?
2. What are three ways that recreational angling can negatively affect fish welfare?
3. The authors reviewed several regulations that are designed to improve fish welfare (Tables 1 and 2 *in* Cooke and Sneddon 2007). They noted the degree of scientific support for each regulation. Which three regulations would be most accepted by anglers? Which three regulations would be the least accepted by anglers?
4. Some countries, such as Germany, ban the practice of catch-and-release angling (all fish captured during angling should be harvested), while other countries such as the United States encourage catch-and-

release angling and base many regulations on this practice (e.g., minimum size limits). What is the future of recreational angling?
   a. From the fish welfare perspective, is the German approach the most progressive, and should all countries adopt this mandate? Alternatively, should catch-and-release angling be encouraged as a way to manage and conserve populations?
   b. Is there room for compromise? In other words, can we allow catch-and-release angling while maintaining fish welfare considerations?
5. Cooke and Sneddon (2007) reviewed several studies that demonstrate negative effects of angling on individual fish. What are potential population level effects of angling? How would these effects alter the dynamics of a population (recruitment, growth, mortality)?

Reference

Cooke, S. J., and L. U. Sneddon. 2007. Animal welfare perspectives on recreational angling. Applied Animal Behaviour Science 104:176–198.

Notes

Notes

Case 31

Rehabilitation Project Assessment for Lentic Habitat Improvement

Background

You have just been hired as a fishery biologist by a state natural resource agency in the Midwest. Your first task is to appear at a public meeting at a town that has a public impoundment located next to it. The local anglers have requested that the agency come to the meeting and provide them with alternatives to improve the fishing in that local impoundment. Long-time residents can remember a time when the fishing was much better than it is now. Your first reaction to this news is to head to the office files. You find that the impoundment is being managed as a centrarchid fishery, with largemouth bass and bluegill two of the primary targets. Spring night electrofishing has occurred annually for nearly three decades. One glance at the sampling data for the last few years quickly confirms that the population abundances for the bass and bluegill are probably quite low, while common carp abundance is quite high. Your next step is to assess the habitat information that has been collected over the years. Again, it does not look like good news—turbid waters (physical turbidity), shallow, wind-swept, and a nearly complete lack of submergent aquatic vegetation.

The three components of a fishery are habitat, biota, and the human dimension. The type, amount, and quality of habitat available will determine the nature of a sport fishery. For example, centrarchid-dominated fisheries often occur in Midwestern lakes where submergent vegetation covers a substantial portion of the lake area. Similarly, the available habitat will determine the suitability of a water body for a rare or nongame fish species.

Habitat management can be an expensive proposition and fishery biologists often do not have access to sufficient funds for large projects. Thus, the Nebraska Game and Parks Commission (NGPC) initiated a habitat improvement program entitled "New Life for Aging Waters." Beginning in 1997, most anglers were required to purchase a $5 aquatic habitat stamp (Figure 31.1). Incidentally, the process to approve the stamp was not easy, as some state legislators viewed the stamp as "one more tax" rather than a "user-fee" requested by that user group. Given annual license sales of approximately

Figure 31.1. The Nebraska Game and Parks Commission initiated a habitat rehabilitation program in 1997 by requiring the purchase of an aquatic habitat stamp.

200,000, the NGPC generated approximately $1 million annually. However, this $1 million is often matched with cooperative efforts. Common examples in Nebraska include Federal Aid in Sport Fish Restoration funds, U.S. Army Corps of Engineers, the Nebraska Environmental Trust Fund, Fish America grants, natural resource districts, and cities. As a result, several million dollars are available most years, which often allow the agency to complete more than one major restoration each year.

Through the normal aging process, impoundments tend to fill with sediment due to runoff from the surrounding watershed or sediment inflow from rivers and streams that feed into the impoundment. This increase in sediment loading can cause changes in water quality, water depth, and habitat complexity. Management efforts typically involve sediment removal and basin sculpturing. In addition, the rehabilitation process provides an ideal time for watershed management efforts designed to decrease erosion and sediment inflow. Such efforts may involve maintenance of filter strips throughout the watershed, or may involve silt check dams constructed above the impoundment. Each project is unique and selected rehabilitation techniques will vary by situation. More information on rehabilitation methods will be provided during your reading assignment for this case study.

Your instructor will present information to you on the rehabilitation efforts at Cottonmill Lake, a 17.4-ha impoundment in Nebraska. You will be asked to determine the likelihood of a successful fish community improvement post-rehabilitation; you will also be asked to determine the economic feasibility of such management efforts. Prior to the rehabilitation, the fish community was dominated by a high biomass of common carp. At the time of construction, the impoundment had a mean depth of 3.6 m; by 1994, mean depth had decreased to 0.6 m. The estimated cost for habitat rehabilitation at Cottonmill Lake was $1.5 million in 1999 dollars (Figure 31.2).

Figure 31.2. Aerial view of Cottonmill Lake, Nebraska in 2003 after the lake rehabilitation in 1999. Note the jetties, islands, and other irregularities that diversify the habitat and create protection from wind and waves in this 17.4-ha impoundment.

Assignment

Read the chapter on lake and reservoir habitat management by Summerfelt (1999).

Questions

1. Would you expect to see any changes in relative abundance indices from standardized sampling for bluegill, channel catfish, and largemouth bass pre- and post-rehabilitation? All three of these fish species were present prior to the rehabilitation, and all three were restocked after the impoundment was drained, habitat was improved, and the lake refilled.
2. What would you expect for physical turbidity measures of water samples taken prior to and after the rehabilitation? Why?
3. Given the high cost of the rehabilitation ($1.5 million), what is the economic feasibility of such a management strategy? Given the high cost of restoration, can we complete this sort of rehabilitation for every aging impoundment? Is this type of restoration feasible in other states that do not have a habitat stamp program like Nebraska? Is it likely that fishery biologists can economically justify the expenditure in terms of cost/benefit?

References

Summerfelt, R. C. 1999. Lake and reservoir habitat management. Pages 285–320 *in* C. C. Kohler and W. A. Hubert, editors. Inland fisheries management in North America, 2nd edition. American Fisheries Society, Bethesda, Maryland.

Notes

Notes

Case 32

Effects of Freshwater Fish Translocations and Stocking Programs, Including the Role of State and Provincial Natural Resource Agencies

Background

The newspaper headline screams: "Biologists stock the wrong subspecies of cutthroat trout into recovery area!" We suspect that none of you would like to wear the shoes of such a biologist. However, this is a real headline that appeared a few years ago in a wide range of newspapers and magazines. The topic of fish stocking can be a hotbed of contention!

For the purposes of this case study, think broadly of fish stocking as:

- fish translocations to water bodies where they are not native;
- fish stocking programs, often for recreational fishing purposes and often with native fish; and
- the use of hatcheries for recovery of endangered fishes.

Assignment

Your instructor will create several teams and then assign specific readings to each team. The teams will develop at least some expertise in several topic areas.

Topic	*Assignment*
1. Interspecific effects on native fishes	*Dunham et al. 2002*
2. Intraspecific effects on native fishes	*Hansen 2002*
3. The role of stocking in recreational fishing	*Heidinger 1999; Lathrop et al. 2002*
4. Concerns over genetics	*Doyle et al. 2001; Philipp et al. 2002*
5. Effects on ecosystems	*Eby et al. 2006; Vitule et al. 2009*

6. Endangered fish recovery	*Anders 1998; Dryer and Sandvol 1993*
7. Agency response to stocking concerns	*Murphy and Kelso 1986; Jackson et al. 2004*

Plan to come to the next class period ready to participate fully in a group discussion. Consider the questions below before that class discussion.

Questions

1. Have fish translocations consistently had "positive" effects or "negative" effects? Why or why not?
2. Do stocking programs for recreational fisheries have consistently positive or negative effects? Why or why not?
3. Are there legitimate genetic concerns over fish translocations, recreational fish stockings, and culture programs for recovery of endangered fishes? If so, what are they?
4. Have the state natural resource agencies ignored the controversy over fish stocking, or have they modified their stocking policies, procedures and practices? What about the future; what changes would you propose?

References

Anders, P. J. 1998. Conservation aquaculture and endangered species: can objective science prevail over risk anxiety? Fisheries 23(11):29–31.

Doyle, R. W., R. Perez-Enriquez, M. Takagi, and N. Taniguchi. 2001. Selective recovery of founder genetic diversity in aquacultural broodstocks and captive, endangered fish populations. Genetica 111:291–304.

Dryer, M. P., and A. J. Sandvol. 1993. Pallid sturgeon, *Scaphirhynchus albus*, recovery plan. U.S. Fish and Wildlife Service, Washington, D.C. Available: http://www.fws.gov/yellowstonerivercoordinator/pallid%20recovery%20plan.pdf (January 2010).

Dunham, J. B., S. B. Adams, R. E. Schroeter, and D. C. Novinger. 2002. Alien invasions in aquatic ecosystems: toward an understanding of brook trout invasions and potential impacts on inland cutthroat trout in western North America. Reviews in Fish Biology and Fisheries 12:373–391.

Eby, L. A., W. J. Roach, L. B. Crowder, and J. A. Stanford. 2006. Effects of stocking-up freshwater food webs. Trends in Ecology and Evolution 21:576–584.

Hansen, M. M. 2002. Estimating the long-term effects of stocking domesticated trout into wild brown trout (*Salmo trutta*) populations: an approach using microsatellite DNA analysis of historical and contemporary samples Molecular Ecology 11:1003–1015.

Heidinger, R. C. 1999. Stocking for sport fisheries enhancement. Pages 375–401 *in* C. C. Kohler and W. A. Hubert, editors. Inland fisheries management in North America, 2nd edition. American Fisheries Society, Bethesda, Maryland.

Jackson, J. R., J. C. Boxrucker, and D. W. Willis. 2004. Trends in agency use of propagated fishes as a management tool in inland fisheries. Pages 121–138 *in* M. J. Nickum, P. M.

Mazik, J. G. Nickum, and D. D. MacKinlay, editors. Propagated fish in resource management. American Fisheries Society, Symposium 44, Bethesda, Maryland.
Lathrop, R. C., B. M. Johnson, T. B. Johnson, S. R. Carpenter, T. R. Hrabik, J. F. Kitchell, J. J. Magnuson, L. G. Rudstam, and R. S. Stewart. 2002. Stocking piscivores to improve fishing and water clarity: a synthesis of the Lake Mendota biomanipulation project. Freshwater Biology 47:2410–2424.
Murphy, B. R., and W. E. Kelso. 1986. Strategies for evaluating fresh-water stocking programs: past practices and future needs. Pages 303–313 *in* R. H. Stroud, editor. Fish culture in fisheries management. Fish Culture Section and Fisheries Management Section of the American Fisheries Society, Bethesda, Maryland.
Philipp, D. P., J. E. Claussen, T. W. Kassler, and J. M. Epifanio. 2002. Mixing stocks of largemouth bass reduces fitness through outbreeding depression. Pages 349–364 *in* D. P. Philipp and M. S. Ridgway, editors. Black bass: ecology, conservation, and management. American Fisheries Society, Symposium 31, Bethesda, Maryland.
Vitule, J. R. S., C. A. Freire, and D. Simberloff. 2009. Introduction of non-native freshwater fish can certainly be bad. Fish and Fisheries 10:98–108.

Notes

Notes

Figure Credits

Title Page

Fossil fish illustration by Diane E. Brown.

Case 1 A Tale of Two Oceans: the Demise of BluefinTuna

Figure 1.1. Bluefin tuna: Drawing provided courtesy of the Maine Department of Marine Resources Recreational Fisheries program and the Maine Outdoor Heritage Fund.

Figure 1.2. Mattanza: Photo courtesy of the NOAA Fisheries Collection: Image ID: fish2013 http://www.photolib.noaa.gov/bigs/fish2013.jpg

Figure 1.3. Tonnara or trap fishing in the Mediterranean: Image courtesy of the NOAA Fish Collection: Image ID: fish2059 http://www.photolib.noaa.gov/htmls/fish2059.htm

Figure 1.4. Purse seine: Illustration by Philip Davis.

Figure 1.5. Longline fishing image courtesy of Australian Fisheries Management Authority http://www.afma.gov.au/information/students/methods/pelagic.htm

Figure 1.6. Frozen tuna photo courtesy of Wikimedia Commons: http://commons.wikimedia.org/wiki/File:Tsukiji_Fish_market_and_Tuna.JPG

Case 2 What's for Dinner?—Environmentally Conscious Seafood Choices

Figure 2.1. Fish illustrations: Drawings provided courtesy of the Maine Department of Maine Resources Recreational Fisheries program and the Maine Outdoor Heritage Fund.

Case 3 ***A Float Trip on the Bad River, South Dakota: How Will Cyprinid Distributions Change from Headwaters to Mouth?***

Figure 3.1. Map of South Dakota: used courtesy of Craig Milewski.

Figure 3.2. Artwork by Maynard Reece. Used courtesy of the Iowa Department of Natural Resources.

Case 5 ***Communism Meets the Tragedy of the Commons: A Fisheries Management Conflict in Rural China***

Figure 5.1. Map file courtesy of Wikimedia Commons. http://commons.wikimedia.org/wiki/File:China_Guizhou.svg

Case 6 ***What Factors are Related to Condition of Flannelmouth Suckers in the Colorado River?***

Figure 6.1. Image courtesy of R. Scott Rogers.

Figure 6.2. Map of the Colorado River used courtesy of Craig Paukert.

Case 7 ***Evaluating the Population Status of Black Sea Bass: One Step at a Time***

Figure 7.7. Drawing provided courtesy of the Maine Department of Marine Resources Recreational Fisheries program and the Maine Outdoor Heritage Fund.

Case 8 ***Predators Eat Prey: Effects of an Inadvertent Introduction of Northern Pike on an Established Fish Community***

Figure 8.1. Image courtesy of Craig Paukert.

Figure 8.4. Image courtesy of T. J. DeBates.

Case 9 ***Misapplication of a Minimum Length Limit for Crappie Populations Could the Mistake Have Been Avoided?***

Figure 9.1. Image of Lake Alvin courtesy of Timothy Bister.

Case 11 To Stock or Not to Stock: That is the Question—for High Elevation Wilderness Lakes

Figure 11.1. Image by Dave Harper, provided courtesy of Hilda Sexauer, Wyoming Game and Fish Department.

Figure 11.2. Image by Bob Wiley, used courtesy of Wyoming Wildlife.

Figure 11.4. Image courtesy of Christopher S. Guy.

Figure 11.5. Image used courtesy of Wyoming Wildlife.

Figure 11.6. Image used courtesy of Wyoming Wildlife.

Figure 11.7. Image courtesy of Christopher S. Guy.

Case 12 A Protected Slot Length Limit for Largemouth Bass in a Small Impoundment: Will the Improved Size Structure Perist?

Figure 12.1. Image used courtesy of Tracy Hill.

Case 13 Horseshoe Crabs: a Struggle among User Groups

Figure 13.1. Image courtesy of Larissa Graham.

Figure 13.2. Image courtesy of Larissa Graham.

Case 15 The Debate Over Shark Abundance

Figure 15.1. Drawings provided courtesy of the Maine Department of Marine Resources Recreational Fisheries program and the Maine Outdoor Heritage Fund.

Case 17 Managing Lake Oahe Walleye in the Face of an Imbalanced Food Web

Figure 17.5. Fish drawings: These images were developed by Ellen Edmonson and Hugh Chrisp for the 1927–1940 New York Biological Survey. Provided courtesy of the New York State Department of Environmental Conservation. All rights reserved. Plankton and macroinvertebrates: Illustrations by Philip Davis.

Case 18 Sea Lions: a New Kind of Nuisance

Figure 18.1. Image courtesy of Byron Grams (bygrams@pacbell.net).

Figure 18.2. Image courtesy of http://www.indospectrum.com.

Figure 18.3. Image courtesy of Gloria Schoenholtz.

Figure 18.4. Image courtesy of Latitude 38 Magazine.

Case 19 Size-Structure Assessment for Pallid Sturgeon

Figure 19.1. Image courtesy of the U.S. Fish and Wildlife Service, Bismarck, ND.

Figure 19.2. Modified from a map by the U.S. Army Corps of Engineers.

Figure 19.3. Modified from a map by the U.S. Army Corps of Engineers.

Case 20 Standardized Sampling: Lake Meredith, Texas

Figure 20.2. Image courtesy of Brian Van Zee.

Figure 20.3. Image courtesy of Brian Van Zee.

Figure 20.4. Image courtesy of Brian Van Zee.

Case 21 Northern Pike Reproduction and Early Life History: Ties to Recruitment Patterns

Figure 21.1. Image used courtesy of the South Dakota Office of Tourism.

Figure 21.3. Image courtesy of John Lott.

Figure 21.4. Image courtesy of John Lott.

Figure 21.5. Image courtesy of John Lott.

Case 22 *Successful Use of a Protected Slot Length Limit to Improve Largemouth Bass Size Structure: But What Happens to the Panfish?*

Figure 22.1. Image used courtesy of the Missouri Natural Resources Conservation Service. http://www.mo.nrcs.usda.gov/news/MOphotogallery/watersheds.html

Figure 22.2. Image courtesy of Steve Eder.

Case 23 *Why Does it Look Like That? How Morphology is Related to Ecology and Management*

Drawings in this case courtesy of Philip Davis.

Case 25 *Dear Old Dad: Fisheries Meets the Stock Market*

Box 25.1. From the August 2005 issue of Motley Fool Hidden Gems. Reprinted by permission from The Motley Fool.

Figure 25.1. Menhaden image courtesy of the Maine Department of Marine Resources Recreational Fisheries program and the Maine Outdoor Heritage Fund.

Case 26 *Exotic Species, Economic Development, and Native Fish Restoration: are All Possible?*

Figure 26.1. Available: http://en.wikipedia.org/wiki/File:Brazil_State_SantaCatarina.svg.

Figure 26.2. Fish images courtesy of LAPAD/Federal University of Santa Catarina, Brazil.

Figure 26.3. Fish images courtesy of LAPAD/Federal University of Santa Catarina, Brazil.

Case 28 *Managing Prey Resources in Colorado Reservoirs*

Figure 28.1. Illustration courtesy of Philip Davis.

Case 29 *Can We Make Overfishing Less Prevalent around the World?*

Figure 29.1. Reprinted by permission from Macmillan Publishers Ltd: [Nature] (Pitcher, T., G. Pramod, D. Kalikoski, and K. Short. 2009. Not honouring the code. Nature 457:658–659.).

Case 31 *Rehabilitation Project Assessment for Lentic Habitat Improvement*

Figure 31.1. Image courtesy of D.W. Gabelhouse, Jr., Nebraska Game and Parks Commission.

Figure 31.2. Image courtesy of B.A. Newcomb, Nebraska Game and Parks Commission.

Figure 31.3. Images courtesy of B.A. Newcomb, Nebraska Game and Parks Commission.

Figure 31.4. Images courtesy of B.A. Newcomb, Nebraska Game and Parks Commission.

Case Index

Numbers refer to corresponding case studies

M

O

P

R

S

W